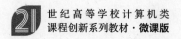

21世纪高等学校计算机类课程创新系列教材·微课版

Android
移动应用设计与开发教程 微课视频版

程杰 / 主编
王天顺 / 副主编

清华大学出版社
北京

内 容 简 介

本书全面介绍Android(安卓)开发技术，内容涵盖了从基础知识到高级应用的各种开发技能。

全书共11章，按Android应用开发知识的系统性，由浅入深地安排内容，全面介绍了Android系统和开发环境、Android常见界面布局与常见界面控件、页面活动单元Activity、多媒体应用开发、数据存储与I/O、使用内容提供者共享数据、广播机制、Service、网络编程及综合项目等内容。每部分内容既有理论知识又有具体实例，通过具体实例将各知识点结合起来，帮助学生理解相关知识，并达到学以致用的目的。本书每章配有小结和习题，便于教学和学习巩固。

本书内容丰富，实例典型，知识讲解系统，适合作为高等学校计算机及相关专业的教材或参考书，也适合软件开发人员及其他有关人员作为自学参考书或培训教材。

版权所有，侵权必究。举报：010-62782989，beiqinquan@tup.tsinghua.edu.cn。

图书在版编目(CIP)数据

Android移动应用设计与开发教程：微课视频版/程杰主编.—北京：清华大学出版社，2024.6(2025.2重印)
21世纪高等学校计算机类课程创新系列教材：微课版
ISBN 978-7-302-66399-7

Ⅰ.①A… Ⅱ.①程… Ⅲ.①移动终端－应用程序－程序设计－高等学校－教材 Ⅳ.①TN929.53

中国国家版本馆CIP数据核字(2024)第111274号

责任编辑：黄　芝　张爱华
封面设计：刘　键
责任校对：郝美丽
责任印制：刘　菲

出版发行：清华大学出版社
　　　网　　址：https://www.tup.com.cn，https://www.wqxuetang.com
　　　地　　址：北京清华大学学研大厦A座　　　邮　编：100084
　　　社 总 机：010-83470000　　　邮　购：010-62786544
　　　投稿与读者服务：010-62776969，c-service@tup.tsinghua.edu.cn
　　　质量反馈：010-62772015，zhiliang@tup.tsinghua.edu.cn
　　　课件下载：https://www.tup.com.cn，010-83470236
印 装 者：三河市人民印务有限公司
经　　销：全国新华书店
开　　本：185mm×260mm　　　印　张：18　　　字　数：437千字
版　　次：2024年8月第1版　　　　　　　　　印　次：2025年2月第2次印刷
印　　数：1501～2500
定　　价：59.80元

产品编号：103662-01

Android 是一种开源的操作系统，由 Google 公司主导并与其他厂商合作开发，广泛应用于移动设备领域。Android 应用程序具有丰富的用户界面和交互设计，可以充分利用设备的硬件和软件资源，提供个性化的用户体验。同时，Android 应用程序还可以与云服务、物联网等技术结合，实现更加丰富的功能和应用场景。

本书以初学者为对象，由浅入深、循序渐进地介绍 Android 移动应用设计与开发的基本概念和基本方法，在内容上力争主次分明，把 Android 技术的一些先进内容和思想方法介绍给读者，希望在有限的篇幅中使读者比较完整地掌握 Android 移动应用设计的思想和方法。

全书共 11 章。第 1 章主要涵盖了 Android 移动应用设计与开发中的基本概念和应用。初学者将会在这部分学习 Android 系统的起源、发展和现状，Android 开发的意义以及应用领域，同时还将学习搭建 Android 开发环境和掌握常用开发工具等。第 2~4 章详细全面地介绍了用户界面设计，包括界面布局、控件使用、自定义控件以及界面交互等。通过这一部分的学习，读者将掌握如何设计一个美观、易用的用户界面。第 5 章介绍了 Android 应用中开发多媒体应用，包括音频、视频、图片等多媒体文件的播放和处理。通过这一章的学习，读者将掌握如何使用 Android 的多媒体框架来开发多媒体应用。第 6 章介绍了如何在 Android 应用中进行基本数据操作以及数据存储的方法，包括 SharedPreferences 存储、File 存储、SQLite 存储等。通过这一章的学习，读者将掌握如何在 Android 应用中处理和保存数据。第 7~9 章介绍了 Android 应用的核心组件，包括内容提供者（ContentProvider）、广播接收者（BroadcastReceiver）和服务（Service）等。通过这一部分的学习，读者将掌握如何使用这些组件来构建一个完整的 Android 应用。第 10 章介绍了如何在 Android 应用中使用网络编程和通信技术，包括 HTTP 请求、Retrofit 框架、JSON 等。通过这一章的学习，读者将掌握如何让自己的应用连接到互联网并与 Web 服务进行通信。第 11 章通过开发一个科学饮食管理系统的实例，将前面所学的知识进行综合运用。通过这一章的实践，读者将更加深入地理解和掌握 Android 应用开发的全过程。

本书由程杰任主编，王天顺任副主编。第 1 章由赵会燕编写；第 2、6、7 由张小峰编写；第 3、5 章由焦阳编写；第 4 章由王文青编写；第 8、9 章由王天顺编写；第 10、11 章由程杰编写。全书由程杰和王天顺统稿，司文智、宋瑞琪、王已卓等人参与了文字整理工作。本书在编写过程中得到了郑州升达经贸管理学院领导和同事的关心与支持。

在编写本书的过程中参考了相关文献，在此向这些文献的作者深表感谢。限于编者水平有限，书中难免有疏漏之处，恳请广大读者批评指正。

程 杰

2024 年 4 月

目 录

下载源码

第1章 Android 系统和开发环境 ………………………………………… 1

  1.1 Android 系统简介 …………………………………………………… 1
      1.1.1 Android 的起源 ………………………………………………… 2
      1.1.2 Android 的发展和前景 ………………………………………… 2
      1.1.3 Android 的系统架构 …………………………………………… 5
  1.2 搭建 Android 开发环境 ……………………………………………… 9
      1.2.1 Android Studio 开发工具 ……………………………………… 9
      1.2.2 Android Studio 的安装 ………………………………………… 9
  1.3 开发第一个 Android 应用 …………………………………………… 14
      1.3.1 创建并运行 HelloWorld 项目 ………………………………… 14
      1.3.2 Android 虚拟机的安装 ………………………………………… 16
      1.3.3 Android 应用项目结构分析 …………………………………… 19
  1.4 小结 …………………………………………………………………… 20
  习题1 …………………………………………………………………… 20

第2章 Android 常见界面布局 …………………………………………… 21

  2.1 界面编程和视图 ……………………………………………………… 22
      2.1.1 视图组件和容器组件 …………………………………………… 22
      2.1.2 使用 XML 布局文件控制 UI 界面 …………………………… 22
      2.1.3 使用代码控制 UI 界面 ………………………………………… 23
  2.2 布局和布局分类 ……………………………………………………… 23
      2.2.1 什么是布局 ……………………………………………………… 23
      2.2.2 LinearLayout …………………………………………………… 25
      2.2.3 TableLayout …………………………………………………… 26
      2.2.4 FrameLayout …………………………………………………… 29
      2.2.5 RelativeLayout ………………………………………………… 31
      2.2.6 GridLayout ……………………………………………………… 33
      2.2.7 AbsoluteLayout ………………………………………………… 37
      2.2.8 ConstraintLayout ……………………………………………… 39
  2.3 小结 …………………………………………………………………… 41
  习题2 …………………………………………………………………… 41

第 3 章 Android 常见界面控件 … 43

3.1 基础控件的使用 … 44
3.1.1 TextView … 44
3.1.2 Button … 44
3.1.3 EditText … 46
3.1.4 ImageView … 47
3.1.5 RadioButton … 47
3.1.6 CheckBox … 48
3.1.7 Toast … 49

3.2 ProgressBar 及其子类 … 50
3.2.1 ProgressBar 的功能和用法 … 50
3.2.2 SeekBar 的功能和用法 … 51
3.2.3 RatingBar 的功能和用法 … 52

3.3 对话框的使用 … 53
3.3.1 使用 AlertDialog 建立对话框 … 54
3.3.2 创建单选和多选对话框 … 54
3.3.3 创建 DatePickerDialog 和 TimePickerDialog 对话框 … 58

3.4 ListView 的使用 … 58
3.4.1 ListView 控件的简单使用 … 58
3.4.2 常用数据适配器 … 59
3.4.3 自定义 ListItem … 59

3.5 RecyclerView 的使用 … 59
3.6 手势 … 61
3.6.1 手势检测 … 62
3.6.2 增加手势 … 62

3.7 应用实例：图片浏览器 … 63
3.8 小结 … 68
习题 3 … 68

第 4 章 页面活动单元 Activity … 69

4.1 创建、注册和使用 Activity … 69
4.1.1 创建 Activity … 70
4.1.2 注册 Activity … 71
4.1.3 使用 Activity … 72
4.1.4 Activity 的启动与关闭 … 73

4.2 Activity 的生命周期和启动模式 … 73
4.2.1 Activity 的生命周期状态 … 73

4.2.2　Activity 的生命周期方法 ·············· 75
　　　4.2.3　Activity 的启动模式 ················ 76
　4.3　Activity 之间的跳转 ······················ 78
　　　4.3.1　Intent ······························ 78
　　　4.3.2　Activity 的数据传递 ················ 81
　　　4.3.3　Activity 数据回传 ·················· 85
　4.4　Fragment 控件 ·························· 86
　　　4.4.1　Fragment 简介 ···················· 86
　　　4.4.2　Fragment 的生命周期 ··············· 86
　　　4.4.3　Fragment 的创建 ··················· 88
　　　4.4.4　Fragment 的应用 ··················· 88
　4.5　应用实例：餐厅点餐 ······················ 89
　4.6　小结 ···································· 96
　习题 4 ······································ 96

第 5 章　多媒体应用开发 ···················· 98

　5.1　音频和视频的播放 ························ 98
　　　5.1.1　使用 MediaPlayer 类播放音频 ········ 99
　　　5.1.2　使用 AudioEffect 类控制音乐特效 ···· 103
　　　5.1.3　使用 VideoView 控件播放视频 ······· 106
　5.2　使用 MediaRecorder 类录制音频 ··········· 108
　5.3　控制摄像头拍照 ·························· 110
　　　5.3.1　通过 Camera 进行拍照 ·············· 110
　　　5.3.2　录制视频短片 ······················ 114
　5.4　应用实例：视频播放器 ···················· 118
　5.5　小结 ···································· 121
　习题 5 ······································ 121

第 6 章　数据存储与 I/O ···················· 122

　6.1　SharedPreferences 存储 ··················· 123
　　　6.1.1　SharedPreferences 的使用 ············ 123
　　　6.1.2　SharedPreferences 的存储位置和格式 ··· 124
　6.2　File 存储 ································ 126
　　　6.2.1　打开应用中数据文件的 I/O 流 ········ 128
　　　6.2.2　读写 SD 卡上的文件 ················ 132
　6.3　SQLite 存储 ····························· 138
　　　6.3.1　SQLiteDatabase 简介 ················ 138
　　　6.3.2　SQLiteOpenHelper 类 ··············· 139

    6.3.3　创建数据库和表 ………………………………………………………… 140
    6.3.4　使用 SQL 语句操作 SQLite 数据库 ………………………………… 140
    6.3.5　使用特定方法操作 SQLite 数据库 ………………………………… 142
    6.3.6　事务 …………………………………………………………………… 144
  6.4　应用实例：手机通讯录 …………………………………………………………… 145
  6.5　小结 ………………………………………………………………………………… 152
  习题 6 …………………………………………………………………………………… 152

第 7 章　使用内容提供者共享数据 ……………………………………………………… 154
  7.1　数据共享标准：ContentProvider …………………………………………………… 154
    7.1.1　ContentProvider 简介 …………………………………………………… 155
    7.1.2　Uri 简介 ………………………………………………………………… 155
    7.1.3　使用 ContentResolver 操作数据 ……………………………………… 157
  7.2　开发 ContentProvider ……………………………………………………………… 158
    7.2.1　开发 ContentProvider 的子类 ………………………………………… 158
    7.2.2　使用 ContentResolver 调用方法 ……………………………………… 160
  7.3　系统的 ContentProvider …………………………………………………………… 161
  7.4　监听 ContentProvider 的数据改变 ………………………………………………… 165
  7.5　应用实例：读取联系人 …………………………………………………………… 173
  7.6　小结 ………………………………………………………………………………… 178
  习题 7 …………………………………………………………………………………… 178

第 8 章　广播机制 ………………………………………………………………………… 180
  8.1　广播机制简介 ……………………………………………………………………… 180
  8.2　发送广播 …………………………………………………………………………… 182
    8.2.1　定义广播接收者 ………………………………………………………… 182
    8.2.2　注册广播接收者 ………………………………………………………… 184
    8.2.3　发送广播步骤 …………………………………………………………… 185
  8.3　有序广播 …………………………………………………………………………… 188
    8.3.1　有序广播和普通广播的区别 …………………………………………… 188
    8.3.2　有序广播的发送与处理流程 …………………………………………… 189
    8.3.3　有序广播实例 …………………………………………………………… 189
  8.4　系统预定义广播 …………………………………………………………………… 192
  8.5　应用实例：通过广播机制判断手机电量状态 …………………………………… 193
  8.6　小结 ………………………………………………………………………………… 198
  习题 8 …………………………………………………………………………………… 199

第 9 章　Service ………………………………………………………………………… 200
  9.1　Service 简介 ………………………………………………………………………… 201
  9.2　Service 的生命周期 ………………………………………………………………… 201
  9.3　启动 Service ………………………………………………………………………… 203

  9.3.1 创建、配置 Service ………………………………………………… 204

  9.3.2 启动和停止 Service ………………………………………………… 207

  9.3.3 绑定 Service ………………………………………………………… 211

 9.4 应用实例：音乐播放器 …………………………………………………… 214

 9.5 小结 ……………………………………………………………………… 219

 习题 9 ………………………………………………………………………… 219

第 10 章　网络编程 ………………………………………………………… 220

 10.1 通过 WebView 控件浏览网页 …………………………………………… 220

 10.2 通过 HTTP 访问网络资源 ……………………………………………… 224

  10.2.1 HTTP 简介 ………………………………………………………… 224

  10.2.2 JSON 解析 ………………………………………………………… 225

  10.2.3 Retrofit 简介 ……………………………………………………… 231

  10.2.4 通过 Retrofit 框架访问 HTTP 网络资源 ………………………… 234

 10.3 应用实例：天气预报 …………………………………………………… 240

 10.4 小结 …………………………………………………………………… 254

 习题 10 ……………………………………………………………………… 254

第 11 章　综合项目——科学饮食管理系统 ………………………………… 255

 11.1 科学饮食管理系统简介 ………………………………………………… 255

 11.2 功能模块设计 ………………………………………………………… 256

 11.3 数据库设计 …………………………………………………………… 256

  11.3.1 数据库实体 ………………………………………………………… 256

  11.3.2 数据库表设计 ……………………………………………………… 257

 11.4 项目界面显示、操作模块的实现 ……………………………………… 259

  11.4.1 页面导航模块 ……………………………………………………… 259

  11.4.2 登录界面模块 ……………………………………………………… 260

  11.4.3 科学饮食管理系统主界面 ………………………………………… 261

  11.4.4 水果营养信息模块 ………………………………………………… 262

  11.4.5 蔬菜营养信息模块 ………………………………………………… 264

  11.4.6 食谱营养信息模块 ………………………………………………… 265

  11.4.7 搜索食物营养信息模块 …………………………………………… 266

  11.4.8 营养饮食信息模块 ………………………………………………… 268

  11.4.9 DRIs 统计查询模块 ……………………………………………… 269

 11.5 科学饮食 Web 服务端的实现 ………………………………………… 272

  11.5.1 保存用户 DRIs 营养素摄入量信息 ……………………………… 272

  11.5.2 统计用户 DRIs 营养素摄入量信息 ……………………………… 273

 11.6 小结 …………………………………………………………………… 274

参考文献 ……………………………………………………………………… 275

第1章 Android系统和开发环境

本章导图

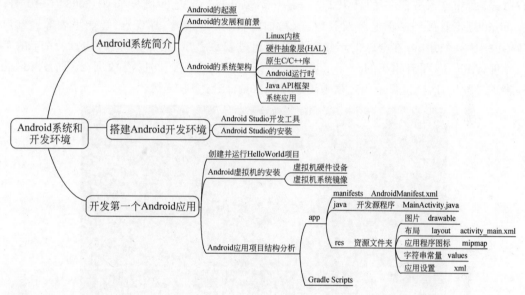

主要内容

- 了解 Android 系统的起源与发展。
- 理解 Android 的系统架构。
- 掌握 Android 应用开发环境的配置。

难点

Android 的系统架构。

本章主要介绍 Android 系统的起源与发展、Android 的系统架构，让读者能够自行下载并安装配置 Android 的开发环境，使用 Android Studio 开发工具编写和运行 Android 应用程序。

1.1 Android 系统简介

Android 是 Google 公司（以下简称 Google）于 2007 年 11 月 5 日宣布的基于 Linux 平

台的开源手机操作系统的名称,该平台由操作系统、中间件、用户界面和应用软件组成。Android 是一种基于 Linux 内核(不包含 GNU 组件)的自由及开放源代码的移动操作系统,主要应用于移动设备,如智能手机和平板电脑,由美国 Google 和开放手机联盟领导及开发。

1.1.1 Android 的起源

Android 一词最先出现在法国作家利尔亚当在 1886 年发表的科幻小说《未来夏娃》中,作者将外表像人类的机器起名为 Android,这也就是 Android 小机器人名字的由来。

Android 操作系统最初由安迪·鲁宾(如图 1.1 所示)开发,主要支持手机。2005 年 8 月由 Google 收购注资。2007 年 11 月,Google 与 84 家硬件制造商、软件开发商及电信营运商组建开放手机联盟共同研发改良 Android 系统。随后 Google 以 Apache 开源许可证的授权方式,发布了 Android 的源代码。第一部 Android 智能手机发布于 2008 年 10 月。Android 逐渐扩展到平板电脑及其他设备,如电视、数码相机、游戏机、智能手表等。2011 年第一季度,Android 在全球的市场份额首次超过塞班系统,跃居全球第一。2013 年的第四季度,Android 平台手机的全球市场份额已经达到 78.1%。2013 年 9 月 24 日,Android 迎来了 5 岁生日,全世界采用这款系统的设备数量已经达到 10 亿台。

图 1.1 安迪·鲁宾与 Android 的 Logo

Android 的 Logo 是由 Ascender 公司设计的,诞生于 2010 年,布洛克绘制了一个简单的机器人,它的躯干就像锡罐的形状,头上还有两根天线。其中的文字使用了 Ascender 公司专门制作的称为 Droid 的字体。Android 是一个全身绿色的机器人,绿色也是 Android 的标志。颜色采用了 PMS 376C 和 RGB 中十六进制的♯A4C639 来绘制,这是 Android 操作系统的品牌象征。

1.1.2 Android 的发展和前景

Android 在正式发行之前,最开始拥有两个内部测试版本,并且以著名的机器人名称来对其进行命名,分别是阿童木(AndroidBeta)和发条机器人(Android 1.0)。后来由于涉及版权问题,Google 将其命名规则变更为用甜点作为系统版本的代号。甜点命名法开始于 Android 1.5 发布时。后继的版本代表的甜点的尺寸越变越大,然后按照 26 个字母顺序:纸杯蛋糕(Cupcake,Android 1.5)、甜甜圈(Donut,Android 1.6)、松饼(Eclair,Android

2.0/2.1)、冻酸奶（Froyo，Android 2.2）、姜饼（Gingerbread，Android 2.3）、蜂巢（Honeycomb，Android 3.0）、冰激凌三明治（Ice Cream Sandwich，Android 4.0）、果冻豆（Jelly Bean、Android 4.1 和 Android 4.2）、奇巧（KitKat，Android 4.4）、棒棒糖（Lollipop，Android 5.0）、棉花糖（Marshmallow，Android 6.0）、牛轧糖（Nougat，Android 7.0）、奥利奥（Oreo，Android 8.0）和派（Pie，Android 9.0），如图 1.2 所示。从 Android 10 开始，Android 不会再按照基于美味零食或甜点的字母顺序命名，而是转换为版本号，就像 Windows 操作系统和 iOS 系统一样。

图 1.2　Android 历代 Logo

2003 年 10 月，安迪·鲁宾等人创建 Android 公司，并组建 Android 团队。

2005 年 8 月 17 日，Google 低调收购了成立仅 22 个月的 Android 及其团队。安迪·鲁宾成为 Google 工程部副总裁，继续负责 Android 项目。

2007 年 11 月 5 日，Google 正式向外界展示了这款名为 Android 的操作系统，并且在这天 Google 宣布建立一个全球性的联盟组织，该组织由 34 家手机制造商、软件开发商、电信运营商以及芯片制造商共同组成，并与 84 家硬件制造商、软件开发商及电信营运商组成开放手持设备联盟（Open Handset Alliance）来共同研发改良 Android 系统，这一联盟将支持 Google 发布的手机操作系统以及应用软件，Google 以 Apache 免费开源许可证的授权方式，发布了 Android 的源代码。

2008 年，在 Google I/O 大会上，Google 提出了 Android HAL 架构图，在同年 8 月 18 日，Android 获得了美国联邦通信委员会（FCC）的批准，在 2008 年 9 月，Google 正式发布了 Android 1.0 系统，这也是 Android 系统最早的版本。

2009 年 4 月，Google 正式推出了 Android 1.5 这款手机，从 Android 1.5 版本开始，Google 开始将 Android 的版本以甜品的名字命名，Android 1.5 命名为 Cupcake（纸杯蛋糕）。该系统与 Android 1.0 相比有了很大的改进。

2009年9月，Google发布了Android 1.6的正式版，并且推出了搭载Android 1.6正式版的手机HTC Hero(G3)，凭借着出色的外观设计以及全新的Android 1.6操作系统，HTC Hero(G3)成为当时全球最受欢迎的手机。Android 1.6也有一个有趣的甜品名称，它被称为Donut(甜甜圈)。

2010年2月，Linux内核开发者Greg Kroah-Hartman将Android的驱动程序从Linux内核"状态树"(Staging Tree)上除去，从此，Android与Linux开发主流将分道扬镳。在同年5月，Google正式发布了Android 2.2操作系统。Google将Android 2.2操作系统命名为Froyo(冻酸奶)。

2010年10月，Google宣布Android系统达到了第一个里程碑，即电子市场上获得官方数字认证的Android应用数量已经达到了10万，Android系统的应用增长非常迅速。在2010年12月，Google正式发布了Android 2.3操作系统Gingerbread(姜饼)。

2011年1月，Google称每日的Android设备新用户数量达到了30万，到2011年7月，这个数字增长到55万，而Android系统设备的用户总数达到了1.35亿，Android系统已经成为智能手机领域占有量最高的系统。

2011年8月2日，Android手机已占据全球智能机市场48%的份额，并在亚太地区市场占据统治地位，终结了Symbian(塞班)系统的霸主地位，跃居全球第一。

2011年9月，Android系统的应用数量已经达到了48万，而在智能手机市场，Android系统的占有率已经达到了43%，继续排在移动操作系统首位。10月19日，Google正式发布全新的Android 4.0操作系统，这款系统被Google命名为Ice Cream Sandwich(冰激凌三明治)。

2012年1月6日，Google Android Market已有10万开发者推出超过40万活跃的应用，且大多数免费。Android Market应用程序商店目录在新年首周周末突破40万基准，距离突破30万应用仅4个月。在2011年早些时候，Android Market从20万增加到30万应用也花了4个月。

2013年11月1日，Android 4.4正式发布。从具体功能上讲，Android 4.4提供了各种实用小功能，新的Android系统更智能，添加更多的Emoji表情图案，UI的改进也更现代，如全新的Hello iOS 7半透明效果。

2014年第一季度，Android平台已占所有移动广告流量来源的42.8%，首次超越iOS。

2014年6月26日，Android 5.0发布，命名Lollipop(棒棒糖)。Android 5.0系统使用一种新的Material Design设计风格。

2015年3月10日，Android 5.1发布，它能够同时支持多张SIM卡，加强了设备保护机制，增强了设备的Wi-Fi能力。

2015年9月30日，Android 6.0发布，命名Marshmallow(棉花糖)。新系统的整体设计风格依然保持扁平化的Material Design风格。Android 6.0对软件体验与运行性能进行了大幅度的优化。据测试，Android 6.0可使设备续航时间提升30%。

2016年8月22日，Android 7.0发布，命名Nougat(牛轧糖)。Android 7.0提供新功能以提升性能、生产效率和安全性，通过新的系统行为测试应用，以节省电量和内存，并充分利用多窗口UI，直接回复通知等功能。

2017年8月22日，Android 8.0发布，命名Oreo(奥利奥)。该系统增加了画中画、通知圆点、通知渠道、自动填充框架、自动调整TextView的大小、可下载字体、自适应图标、快捷方式固定、广色域色彩、WebView、Java 8 API、媒体、多显示器支持等功能。

2018年5月9日，Android 9.0发布，命名Pie(派)。Android 9.0利用人工智能技术,让手机可以为用户提供更多帮助。手机变得更智能、更快，并且还可以随着人们的使用进行调整。

2019年9月4日，Android 10发布，从Android 10开始，Google开始提供系统级的黑暗模式，大部分预装应用、抽屉、设置菜单和Google Feed资讯流等界面和按钮，都会变成以黑色为主色调。为确保用户隐私和安全，推出新的保护措施。借助高性能编解码器、更出色的生物识别技术、更快的应用启动速度、Vulkan 1.1、NNAPI 1.2、可折叠设备和5G等，扩展了更多功能扩展。

2020年9月9日，Android 11发布。Android 11主要提升了聊天气泡、安全隐私、电源菜单功能，新增链接KPI，并支持瀑布屏、折叠屏、双屏等功能。

2021年10月5日，Android 12发布。Android 12优化了触发问题，双击背面手势可以截取屏幕截图、召唤Google Assistant、打开通知栏、控制媒体播放或打开最近的应用程序列表。

2022年8月16日，Android 13发布。Android 13系统将全力推动eSIM技术发展，或将支持一颗芯片原生开通多张eSIM虚拟手机卡。Android 13将支持在锁屏界面添加二维码扫描器，可以更方便地扫描二维码。

2022年9月，Android 14发布。Android 14将支持卫星通信技术。

Android平台的诞生为手机智能化的普及立下汗马功劳，但Android平台最大的缺点也越来越凸显，那就是碎片化严重：设备繁多，品牌众多，版本各异，分辨率不统一，等等，这些都逐渐成为Android系统发展的障碍。碎片化严重不仅造成Android系统混乱，也导致Android应用的隐性开发成本的增多。任何成功的智能操作系统都是由庞大的软件资源支撑起来的，这要求系统和硬件有一定的一致性，这才能确保软件的兼容性，而个人和团体开发的第三方软件也有一定的规范，以确保软件和设备完全兼容。而由于Android完全免费以及完全开源的性质，最终导致Android设备的软件兼容性变差，间接加大了软件开发的难度(主要难度是让软件在更多的设备上运行)，最终会导致一个结果：由于开发难度高，因此开发成本增大。

从Android 10开始，Google引入了Project Mainline，将系统功能模块化，并把Android的12个核心组件，也就是媒体编解码器、Conscrypt、权限控制器、模块元数据等模块化。在Android 11中Google将系统内核进行了模块化修改，内核被分成了通用内核镜像(Generic Kernel Image，GKI)和其他GKI模块，特定硬件的驱动程序(可能是闭源驱动)则作为内核模块加载，从而提供了一个稳定的写入接口，使得硬件供应商能够轻松地插入代码，并最终消除特定的设备内核。

Android在我国的前景十分广阔，首先是有成熟的消费者，在国内，Android社区十分红火，这些社区为Android在中国的普及做了很好的推广作用。国内厂商和运营商也纷纷加入了Android阵营，包括中国移动、中国联通、中兴通讯、华为通讯、联想等大企业，同时不仅仅局限于手机，国内厂家也陆续推出了采用Android系统的MID产品。可以预见，Android也将会被广泛应用在国产智能上网设备上，将进一步扩大Android系统的应用范围。

1.1.3 Android的系统架构

虽然Android系统非常庞大且错综复杂，需要具备全面的技术栈，但整体架构设计清晰。Android底层内核空间以Linux内核作为基石，上层用户空间由原生系统库、虚拟机运行环境、框架层组成，通过系统调用(Syscall)连通系统的内核空间与用户空间。对于用户空间主要采用C++和Java语言编写代码，通过JNI技术打通用户空间的Java层和原生层(C++/C)，从而连通整个系统。

　　Android的系统架构从下往上依次分为Linux内核(Linux Kernel)、硬件抽象层(HAL)、原生C/C++库(Native C/C++ Libraries)和Android运行时(Android Runtime)、Java API框架以及应用这5层架构，其中每一层都包含大量的子模块或子系统，其系统架构如图1.3所示。

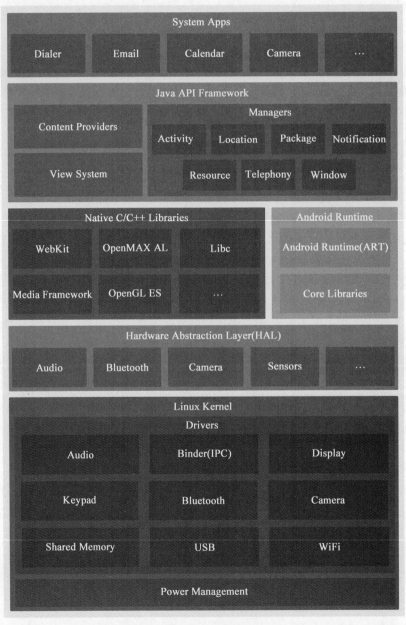

图1.3　Android系统架构

1. Linux 内核

Android 平台的基础是 Linux 内核。Android 核心系统服务依赖于 Linux 内核,包括安全性、内存管理、进程管理、网络协议和驱动模型等。Linux 内核也同时作为硬件和软件栈之间的抽象层。例如 Android 运行时(ART)依靠 Linux 内核来执行底层功能,例如线程和底层内存管理。使用 Linux 内核可让 Android 利用主要安全功能,并且允许设备制造商为著名的内核开发硬件驱动程序。

2. 硬件抽象层(HAL)

硬件抽象层介于 Linux 内核和系统运行层,它是对下层 Linux 驱动的统一封装,并且对上层提供接口,隐藏了底层的细节。它运行在用户空间(User Space),这样的一层就是专门为各个厂商服务的,方便其设计自己的风格。硬件抽象层提供标准界面,向更高级别的 Java API 框架显示设备硬件功能。硬件抽象层包含多个库模块,其中每个模块都为特定类型的硬件组件实现一个界面,例如相机或蓝牙模块。当框架 API 要求访问设备硬件时,Android 系统将为该硬件组件加载库模块。

3. 原生 C/C++ 库

许多核心 Android 系统组件和服务(例如 ART 和 HAL)构建自原生代码,需要以 C 和 C++编写的原生库。在 Linux 内核之上,Android 提供了各种 C/C++核心库(例如 Libc 和 SSL)、视频音频相关的媒体库、外观管理器。基于 2D、3D 图形 SGL 和 OpenGL 图形库、用于本地数据库支持的 SQLite,以及用于集成 Web 浏览器和 Internet 安全的 SSL 和 WebKit,如表 1.1 所示。Android 平台提供 Java 框架 API 以向应用显示其中部分原生库的功能。例如,可以通过 Android 框架的 Java OpenGL API 访问 OpenGL ES,以支持在应用中绘制和操作 2D 和 3D 图形。如果开发的是需要 C 或 C++代码的应用,可以使用 Android NDK 直接从原生代码访问某些原生平台库。

表 1.1　C/C++ 程序库

名　　称	功　能　描　述
OpenGL ES	3D 绘图函数库
Libc	从 BSD 继承来的标准 C 系统函数库,专门为基于嵌入式 Linux 的设备定制
Media Framework	多媒体库,支持多种常用的音频、视频格式录制和回放
SQLite	轻型的关系数据库引擎
SGL	底层的 2D 图形渲染引擎
SSL	安全套接层,是为网络通信提供安全及数据完整性的一种安全协议
FreeType	可移植的字体引擎,它提供统一的接口来访问多种字体格式文件

4. Android 运行时

对于运行 Android 5.0(API 级别为 21)或更高版本的设备,每个应用都在其自己的进程中运行,并且有其自己的 ART 实例。ART 编写为通过执行 DEX(Dalvik Executable)文件在低内存设备上运行的多个虚拟机,DEX 文件是一种专为 Android 设计的字节码格式,经过优化,使用的内存很少。编译工具链(例如 Jack)将 Java 源代码编译为 DEX 字节码,使其可在 Android 平台上运行。

ART 的部分主要功能包括:

(1) 预先(AOT)和即时(JIT)编译。

(2) 优化的垃圾回收(Garbage Collection,GC)。

(3) 在 Android 9(API 级别为 28)及更高版本的系统中,支持将应用软件包中的(DEX)文件转换为更紧凑的机器代码。

(4) 更好的调试支持,包括专用采样分析器、详细的诊断异常和崩溃报告,并且能够设置观察点以监控特定字段。

可以让一个 Android 手机从本质上与一个移动 Linux 实现区分开来。由于 Android 运行时包含了核心库和 Dalvik 虚拟机,因此 Android 运行时是向应用程序提供动力的引擎,并与之一起形成了应用程序框架的基础。其中,Android 库提供了 Java 核心库和 Android 特定库的大部分功能;Dalvik 虚拟机是一个基于寄存器的 Java 虚拟机,并对其优化从而确保同一设备可以高效地运行多个实例,通过 Linux 内核对线程和底层内存进行管理。在 Android 版本 5.0(API 级别为 21)之前,Dalvik 是 Android 运行时。如果用户的应用在 ART 上运行结果很好,那么它应该也可在 Dalvik 上运行,但反过来不一定。

5. Java API 框架

可通过以 Java 语言编写的 API 使用 Android 系统的整个功能集。这些 API 形成创建 Android 应用所需的构建块,它们可简化核心模块化系统组件和服务的重复使用,包括以下组件和服务:

(1) 丰富、可扩展的视图系统,可用以构建应用的 UI,包括列表、网格、文本框、按钮甚至可嵌入的网络浏览器。

(2) 资源管理器,用于访问非代码资源,例如本地化的字符串、图形和布局文件。

(3) 通知管理器,可让所有应用在状态栏中显示自定义提醒。

(4) Activity 管理器,用于管理应用的生命周期,提供常见的导航返回栈。

(5) 内容提供程序,可让应用访问其他应用(例如"联系人"应用)中的数据或者共享其自己的数据。

开发者可以完全访问 Android 系统应用使用的框架 API。该层的编写核心便是 API 框架,是 Android 为开发者提供的开发平台,其也是 Android 平台整体的核心机制,主要组件如表 1.2 所示。

表 1.2 Java API 框架主要组件

名 称	功 能 描 述
Activity Manager(活动管理器)	管理各个应用程序生命周期,以及常用的导航回退功能
Location Manager(位置管理器)	提供地理位置及定位功能服务
Package Manager(包管理器)	管理所有安装在 Android 系统中的应用程序
Notification Manager(通知管理器)	使得应用程序可以在状态栏中显示自定义的提示信息
Resource Manager(资源管理器)	提供应用程序使用的各种非代码资源,如本地化字符串、图片、布局文件、颜色文件等
Telephone Manager(电话管理器)	管理所有的移动设备功能
Window Manager(窗口管理器)	管理所有开启的窗口程序
Content Provider(内容提供者)	使得不同应用程序之间可以共享数据
View System(视图系统)	构建应用程序的基本组件

6. 系统应用

Android 随附一套用于电子邮件、短信、日历、互联网浏览和联系人等的核心应用。平台随附的应用与用户可以选择安装的应用一样，没有特殊状态。因此，第三方应用可成为用户的默认网络浏览器、短信收发程序甚至默认键盘(有一些例外，例如系统的"设置"应用)。

系统应用可用作用户的应用，以及提供开发者可从其自己的应用访问的主要功能。例如，如果您的应用要发短信，您无须自己构建该功能，可以改为调用已安装的短信应用向您指定的接收者发送消息。

1.2 搭建 Android 开发环境

工欲善其事，必先利其器。在 Android 项目的开发中，借助工具能使开发效率大幅提升。

1.2.1 Android Studio 开发工具

Android Studio 是 Google 推出的一个 Android 集成开发工具，基于 IntelliJ IDEA。类似 Eclipse ADT，Android Studio 提供了集成的 Android 开发工具用于开发和调试，还提供更多可提高 Android 应用构建效率的功能，例如：

(1) 基于 Gradle 的灵活构建系统。
(2) 快速且功能丰富的模拟器。
(3) 统一的环境(供用户开发适用于所有 Android 设备的应用)。
(4) 通过实时编辑功能，实时更新模拟器和实体设备中的可组合项。
(5) 代码模板和 GitHub 集成，可协助用户打造常见的应用功能及导入示例代码。
(6) 大量的测试工具和框架。
(7) Lint 工具，能够找出性能、易用性和版本兼容性等方面的问题。
(8) C++ 和 NDK 支持。
(9) 内置对 Google 云平台的支持，可轻松集成 Google 云消息和 App 引擎。

1.2.2 Android Studio 的安装

Android Studio 是可以从官方网站 https://developer.android.google.cn/?hl=zh-cn 下载最新版本的。如果是在 Windows 上安装 Android Studio，找到名为 android-studio-2022.2.1.20-windows.exe 的文件下载，并通过 Android Studio 向导运行。

Windows 的系统要求如表 1.3 所示。

表 1.3　Windows 的系统要求

要　　求	最　小　值	推　荐
OS	64 位 Microsoft Windows 8	最新的 64 位版 Windows
RAM	8GB RAM	至少有 16GB 的 RAM
CPU	x86_64 CPU 架构；第 2 代 Intel Core 或更高版本，或者支持 Windows Hypervisor Framework 的 AMD CPU	最新的 Intel Core 处理器

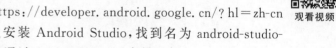
观看视频

续表

要　　求	最　小　值	推　　荐
磁盘可用空间	8GB(IDE 和 Android SDK 及模拟器)	具有 16GB 或更大容量的固态硬盘
屏幕分辨率	1280×800 像素	1920×1080 像素

运行下载好的 Android Studio 安装文件,首先会出现 Android Studio 安装的欢迎界面,如图 1.4 所示。

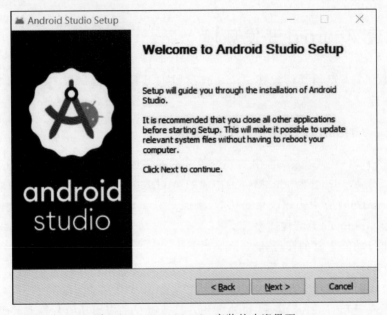

图 1.4　Android Studio 安装的欢迎界面

单击 Next 按钮,进入选择安装组件的界面,这里开发工具 Android Studio 以及 Android 虚拟机(Android Virtual Device)这两个组件都需要安装,如图 1.5 所示。

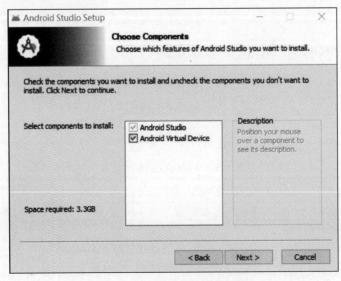

图 1.5　选择安装组件的界面

单击Next按钮,设置安装的位置,如图1.6所示。

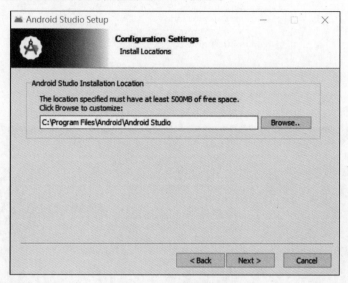

图1.6　设置安装的位置

单击Next按钮开始正式安装,这时会在线下载安装所需要的文件和类库,所需时间较长,需耐心等待,待安装结束后会出现安装完成的界面,如图1.7所示。

图1.7　安装完成的界面

安装完成后,第一次运行Android Studio会弹出对话框询问是否要导入以前的安装配置,这里如果是第一次安装则选择不导入即Do not import settings,如果以前安装过该开发工具并想使用以前的配置则可指定以前的配置文件路径进行导入,如图1.8所示。

然后出现Android Studio运行配置向导,单击Next按钮进入配置类型选项界面,这里可以直接选择标准(Standard)配置,如图1.9所示。

单击Next按钮进入主题界面风格的选择界面,可根据喜好选择黑暗或明亮的主题界面风格,如图1.10所示。

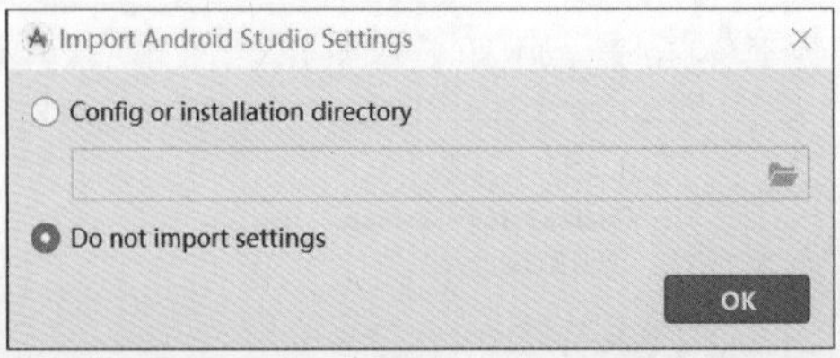

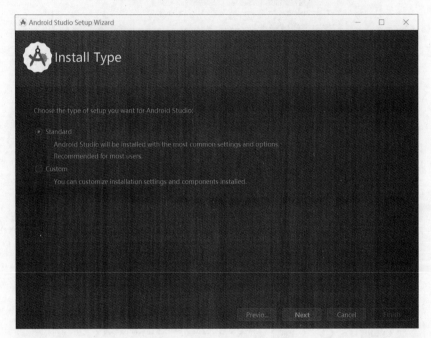

图 1.8 导入 Android Studio 配置对话框

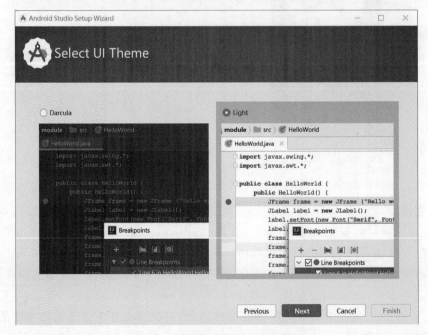

图 1.9 配置类型选择界面

图 1.10 主题界面风格的选择界面

继续单击 Next 按钮，确认所需要安装的类型、路径、SDK 等配置信息后开始安装，如图 1.11 所示。

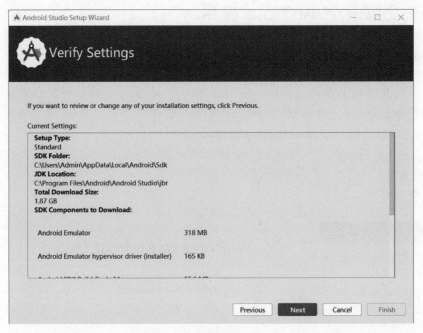

图 1.11　确认安装配置信息

接下来对两个安装授权信息单击 Agree 按钮后开始安装，这里同样需要在线下载相应的安装开发的 SDK 类库，以及 Android 虚拟机所需要的文件，请务必保证网络质量进行下载安装，时间也会稍长，需耐心等待。待成功安装后会出现安装成功的界面，如图 1.12 所示。

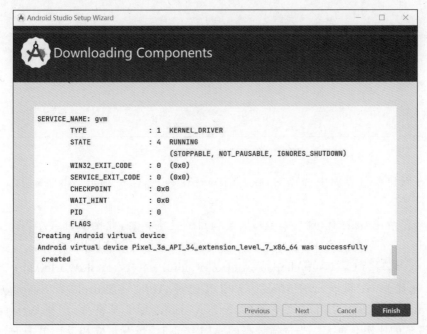

图 1.12　安装成功的界面

安装成功后就可以使用 Android Studio 开发工具进行 Android 项目的开发了。

1.3 开发第一个 Android 应用

1.3.1 创建并运行 HelloWorld 项目

下面通过介绍最简单的 Android 应用程序例子，来了解它的结构和开发流程。例子 HelloWorldApp 给出了一个简单 Android 应用程序创建的过程。首先打开 Android Studio 开发工具，在欢迎界面单击 New Project 按钮，如图 1.13 所示。

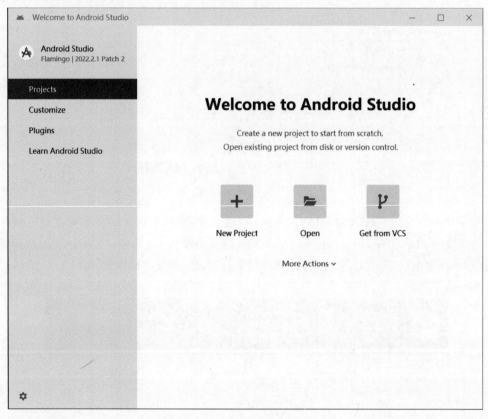

图 1.13　在欢迎界面中新建项目

接着在新建项目界面中选择 Empty Views Activity，然后单击 Next 按钮，如图 1.14 所示。

接下来在新建项目的信息页中填写所建项目的基本信息，此例项目名称 Name 填写为 HelloWorldApp；项目包名 Package name 填写为 com.example.helloworldapp；存储位置 Save location 填写为 C:\Users\bob50\AndroidStudioProjects\HelloWorldApp；开发语言 Language 选择 Java；最小 SDK Minimum SDK 选择 API 24：Android 7.0(Nougat)，当选择较低的 API 级别时，应用可以使用的现代 Android API 会更少，但能够运行应用的 Android 设备的比例会更大。当选择较高的 API 级别时，情况正好相反。单击 Finish 按钮开始创建项目，如图 1.15 所示。

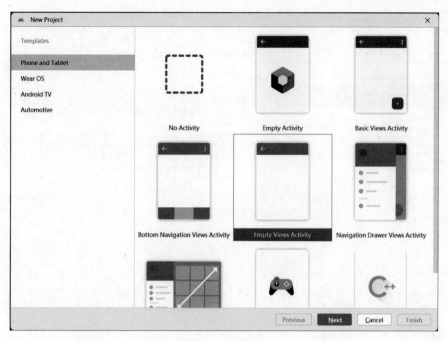

图 1.14　新建项目

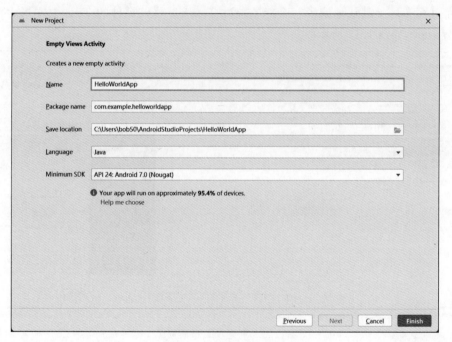

图 1.15　新建项目信息

默认情况下，Android Studio 会将项目配置为使用 AndroidX 库，AndroidX 库需要将编译 SDK 设置为 Android 9.0（API 级别为 28）或更高版本。Android Studio 会在创建新项目时加入一些基本的代码和资源，这里需要在线下载所需的资源文件以及项目管理工具 Gradle 等，所需时间较长，需耐心等待。项目创建成功后会出现项目结构和 MainActivity 的代码，如图 1.16 所示。

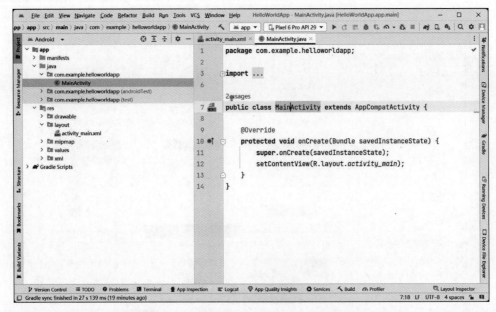

图 1.16　项目结构和 MainActivity 的代码

新建项目的布局资源文件为 activity_main.xml，默认的页面布局为 ConstraintLayout 布局，在页面布局上默认地给了一个标签控件 TextView，该控件的 Text 文本属性的值为 Hello World!，如图 1.17 所示。

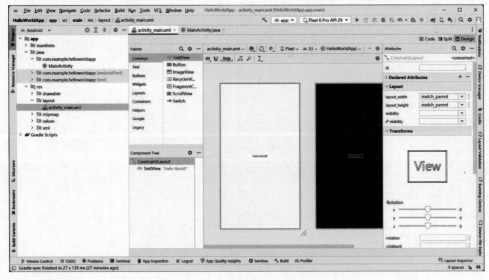

图 1.17　新建项目布局资源

1.3.2　Android 虚拟机的安装

Android 项目创建成功后，如果要运行 Android 应用程序，就需要安装一个 Android 虚拟机或使用 Android 系统的手机来进行运行。但真机运行需要使用数据线与开发计算机相连并需要打开手机系统中的开发者选项，在开发者选项中需要启用 USB 调试功能，每次进行这样的操作过于烦琐,故推荐使用 Android 虚拟机进行运行调试,效果同真机运行调试一致。下面是安装 Android 虚拟机的过程。

在 Android Studio 开发工具的工具栏单击 Device Manager 按钮,如图 1.18 所示。

图 1.18　单击 Device Manager 按钮

在 Device Manager 界面中单击 Create device 按钮,如图 1.19 所示。

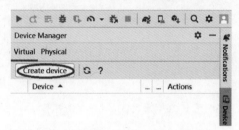

图 1.19　单击 Create device 按钮

接下来选择安装 Android 系统的虚拟机硬件设备,作者在此处选取型号为 Pixel 6 Pro 的手机硬件,如图 1.20 所示。

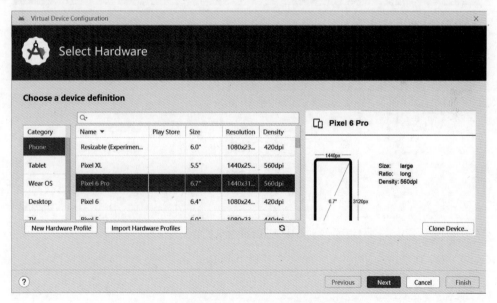

图 1.20　选择 Android 系统的虚拟机硬件设备

单击 Next 按钮后,选取虚拟机需要安装的 Android 系统镜像,因为开发使用的计算机均为 x86 架构的计算机,故此处单击 x86 Images 标签页,在此界面下选取 x86 架构 64 位的 Android 10.0 系统镜像进行下载安装。单击 Next 按钮继续,如图 1.21 所示。

确认 Android 虚拟机的配置信息后单击 Finish 按钮完成安装,如图 1.22 所示。

完成 Android 虚拟机的配置安装后,就可以在 Android Studio 工具栏的可用设备中看

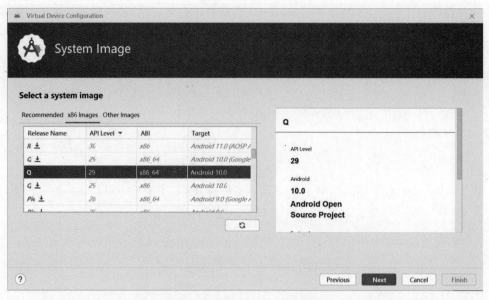

图 1.21　选取 Android 系统镜像

图 1.22　确认 Android 虚拟机的配置信息

到 Pixel 6 Pro API 29 虚拟设备了。单击工具栏上的 Run'app'按钮进行运行调试，如图 1.23 所示。

图 1.23　运行应用程序

Android Studio 会自动运行前面安装的虚拟设备,并在虚拟设备上运行刚刚开发的应用程序 HelloWorldApp,其运行结果如图 1.24 所示。

图 1.24　运行应用程序 HelloWorldApp 结果

1.3.3　Android 应用项目结构分析

默认情况下,Android Studio 会在 Android 视图中显示项目文件结构。此视图并未反映磁盘上的实际文件层次结构。相反,它按模块和文件类型进行整理,以简化项目的关键源文件之间的导航方式,并隐藏某些不常用的文件或目录。

Android 视图与磁盘上的结构之间的一些结构差异在于 Android 视图:

(1) 在顶级 Gradle Script 组中显示项目的所有与构建相关的配置文件。

(2) 在模块级组中显示每个模块的所有清单文件(当针对不同的产品变种和 build 类型使用不同的清单文件时)。

(3) 在一个组(而非在每个资源限定符的单独文件夹)中显示所有备用资源文件。例如,启动器图标的所有密度版本都可以并排显示。

Android 应用项目结构中,文件显示在以下组中:

(1) manifests:包含 AndroidManifest.xml 文件。

(2) java:包含 Kotlin 和 Java 源代码文件(以软件包名称分隔各文件,包括 JUnit 测试代码。

（3）res：包含所有非代码资源（例如 XML 布局、界面字符串和位图图像），这些资源划分到相应的子目录中。

Android 应用项目结构如图 1.25 所示。

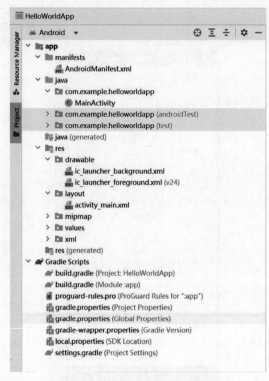

图 1.25　Android 应用项目结构

1.4　小结

本章主要讲解了 Android 的基础知识，首先介绍了 Android 的发展历史及系统架构，然后讲解 Android 开发环境的搭建，接着开发了一个 HelloWorld 程序，了解 Android 项目的搭建、程序的结构，以及资源文件。通过本章的学习，希望读者能对 Android 系统有一个大致的了解，并能够独立搭建 Android 的开发环境，了解 Android 项目的程序结构，为后续学习 Android 知识做好铺垫。

习题 1

1. 简述 Android 系统的发展历程。
2. 简述 Android 系统的系统架构。
3. 简述 Android 开发环境的搭建过程。
4. 任意创建一个 Android 项目并运行。

第 2 章

Android常见界面布局

本章导图

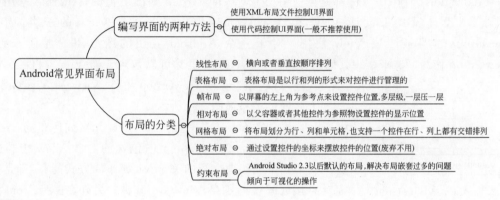

主要内容

- 界面编程和视图。
- 视图组件和容器组件。
- UI界面的控制方式。
- 布局和布局分类。
- 线性布局。
- 表格布局。
- 帧布局。
- 相对布局。
- 网格布局。
- 绝对布局。
- 约束布局。

难点

- UI界面的控制方法。
- 不同布局的特点及使用。

Android系统应用程序一般由多个Activity组成，这些Activity以界面的形式展现在用户面前，界面是由布局和控件组成的，针对界面中控件不同的排列位置，Android定义了相应的布局进行管理。本章将针对 Android 中的界面相关知识及常用的布局进行详细的讲解。

2.1 界面编程和视图

2.1.1 视图组件和容器组件

　　Android 应用的绝大部分 UI 组件都放在 android.widget 包及其子包、android.view 包及其子包中，Android 应用的所有 UI 组件都继承了 View 类，它代表一个空白的矩形区域。View 类还有一个重要的子类：ViewGroup，但 ViewGroup 通常作为其他组件的容器使用。

　　Android 的所有 UI 组件都是建立在 View、ViewGroup 基础之上的，Android 采用了"组合器"设计模式来设计 View 和 ViewGroup：ViewGroup 是 View 的子类，因此 ViewGroup 也可被当成 View 使用。对于一个 Android 应用的图形用户界面来说，ViewGroup 作为容器来盛装其他组件，而 ViewGroup 里除可以包含普通 View 组件之外，还可以再次包含 ViewGroup 组件，View 和 ViewGroup 之间的关系如图 2.1 所示。

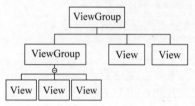

图 2.1　View 和 ViewGroup 之间的关系

　　一般布局方式是指一组 View 元素如何布局，准确地说是一个 ViewGroup 中包含的一些 View 怎样布局。下面小节介绍的关于 View 布局方式的类，都是直接或间接继承自 ViewGroup 类。另外，在 Android 程序中常用的界面布局有两种方法：一种是使用 XML 布局文件控制 UI 界面；另一种是使用代码控制 UI 界面。

2.1.2 使用 XML 布局文件控制 UI 界面

　　在 Android 应用中可以使用 XML 布局文件来控制 UI 界面，它使用一种层次结构的方式描述了界面中各个元素的位置和关系，从而有效地将界面中布局的代码和 Java 代码隔离，使程序的结构更加清晰。因此，在 Android 应用开发中常采用这种方法。

　　通常布局文件都是存放在 Android 应用的 res/layout 文件夹中，用户可以在该文件夹的 XML 文件中编写布局，下面给出了一个 activity_main.xml 的布局示例代码。

```
<LinearLayout xmlns:android = "http://schemas.android.com/apk/res/android"
    android:layout_width = "match_parent"
    android:layout_height = "match_parent"
    android:orientation = "vertical">

    <TextView
        android:layout_width = "wrap_content"
        android:layout_height = "wrap_content"
        android:text = "Hello, World!" />

    <Button
        android:layout_width = "wrap_content"
        android:layout_height = "wrap_content"
```

```
            android:text = "Click Me" />
</LinearLayout>
```

上述代码中,定义了一个线性布局 LinearLayout,它具有垂直方向的排列,在该布局中定义了一个 TextView 和 Button 控件。LinearLayout 继承自 ViewGroup,TextView 继承自 View,Button 继承自 TextView。

2.1.3 使用代码控制 UI 界面

Android 应用的 UI 界面不仅可以使用 XML 进行控制,还可以在 Java 代码中进行控制。在 Android 应用中所有的布局和控件的对象都可以通过 new 关键字创建出来,将创建的 View 控件添加到 ViewGroup 布局中,从而实现 View 控件在 UI 界面中显示。如将 2.1.2 节中的 XML 控制的界面改写成代码控制的 UI 界面,示例代码如下。

```
LinearLayout layout = new LinearLayout(this);
layout.setOrientation(LinearLayout.VERTICAL);      //设置布局垂直排列
TextView textView = new TextView(this);            //创建 TextView 控件
textView.setText("Hello, World!");                 //设置 TextView 控件显示的内容
Button button = new Button(this);                  //创建 Button 控件
button.setHeight(100);
button.setWidth(50);
button.setText("Click Me");                        //设置 Button 控件显示的内容
                                                   //添加 textView 和 button 对象
layout.addView(textView);
layout.addView(button);
setContentView(layout);                            //设置 Activity 中显示
```

在上述代码中首先创建线性布局对象 layout 并使用 setOrientation()方法设置其垂直排列;接着创建了一个文本对象 textView,使用 setText()方法设置其显示的内容以及创建按钮对象 button,使用 setHeight()、setWidth()、setText()方法设置按钮的高度、宽度、显示的内容;最后使用 addView()方法将控件添加到布局中并在 Activity 中显示。

2.2 布局和布局分类

在 Android 应用中会见到各种各样的界面设计,例如 QQ、微信、新浪微博等,为了适应各种各样的界面风格,Android 系统为开发者提供了常用的 7 种布局,分别是线性布局(LinearLayout)、表格布局(TableLayout)、帧布局(FrameLayout)、相对布局(RelativeLayout)、网格布局(GridLayout)、绝对布局(AbsoluteLayout)以及约束布局(ConstraintLayout)。接下来,本节将对这些布局进行重点讲解。

2.2.1 什么是布局

布局就是把界面中的控件按照某种规律摆放在指定的位置,形成不同的程序界面,也可以理解为 Android 应用程序的桌面。它主要是为了解决应用程序在不同手机中的显示问题。Android 系统中常用的布局有 7 种,它们都直接或者间接继承 ViewGroup,因此这 7 种

布局也支持 ViewGroup 中定义属性,也被称为通用属性,具体如表 2.1 所示。

表 2.1 布局的通用属性

属 性 名 称	功 能 描 述
android:id	设置布局的标识
android:layout_width	设置布局的宽度
android:layout_height	设置布局的高度
android:background	设置布局的背景
android:layout_margin	设置当前布局与屏幕边界或与周围控件的距离
android:padding	设置当前布局与该布局中控件的距离

接下来就对表 2.1 中的属性进行详细的说明。

1. android:id

android:id 用于设置当前布局的唯一标识。在 XML 文件中它的属性值是通过"@＋id/属性名称"定义的。为布局指定 android:id 属性后,在 R.java 文件中,会自动生成对应的 int 值。在 Java 代码中通过为 findViewById() 方法传入该 int 值来获取该布局对象。

2. android:layout_width

android:layout_width 用于设置布局的宽度,其值可以是具体的尺寸,如 50dp,也可以是系统定义的值,具体如下:

(1) fill_parent:表示该布局的宽度与父容器(从根元素讲是屏幕)的宽度相同。

(2) match_parent:与 fill_parent 的作用相同,从 Android 2.2 开始推荐使用 match_parent。

(3) wrap_content:表示该布局的宽度恰好能包裹它的内容。

3. android:layout_height

android:layout_height 用于设置布局的高度,其值可以是具体的尺寸,如 50dp,也可以是系统定义的值,具体如下:

(1) fill_parent:表示该布局的高度与父容器的高度相同。

(2) match_parent:与 fill_parent 的作用相同,从 Android 2.2 开始推荐使用 match_parent。

(3) wrap_content:表示该布局的高度恰好能包裹它的内容。

4. android:background

android:background 用于设置布局背景,其值可以引用图片资源,也可以是颜色资源。

5. android:layout_margin

android:layout_margin 用于设置当前布局与屏幕边界、周围布局或控件的距离。属性值为具体的尺寸,如 45dp。与之相似的属性还有 android:layout_marginTop、android:layout_marginBottom、android:layout_marginLeft、android:layout_marginRight,分别用于设置当前布局与屏幕、周围布局或者控件的上、下、左、右边界的距离。

6. android:padding

android:padding 用于设置当前布局内控件与该布局的距离,其值可以是具体的尺寸,如 45dp。与之相似的属性还有 android:paddingTop、android:paddingBottom、android:paddingLeft、android:paddingRight,分别用于设置当前布局中控件与该布局上、下、左、右的距离。

需要注意的是，Android 系统提供的七种常用布局必须设置 android:layout_width 和 android:layout_height 属性指定其宽和高，其他属性可以根据需求进行设置。

2.2.2　LinearLayout

LinearLayout 是一种常用的界面布局方式，在 LinearLayout 中，所有的子元素都按照垂直或水平的顺序在界面上排列，每一个子元素都位于前一个子元素的后面，当超过边界时，将部分显示或者不显示。在 XML 布局文件中定义 LinearLayout 的基本语法格式如下：

```
<LinearLayout xmlns:android = "http://schemas.android.com/apk/res/android"
        属性 = "属性值"
        … …>
</LinearLayout>
```

除了布局的通用属性外，LinearLayout 还有其他常用的属性，具体如表 2.2 所示。

表 2.2　LinearLayout 常用的属性

属 性 名 称	功 能 描 述
android:orientation	设置布局内控件的排列顺序
android:layout_gravity	设置控件在容器中的对齐方式
android:layout_weight	在布局内设置控件权重，属性值可直接写 int 值

1. 属性说明

针对表 2.2 中的属性进行详细的讲解，具体如下：

（1）android:orientation：指定布局控件的排列方式。取值为 vertical 时，表示从上至下垂直排列；取值为 horizontal 时，表示从左到右水平排列。

（2）android:layout_gravity：指定布局控件在容器中的对齐方式。

（3）android:layout_weight：指定 View 控件在容器中所占的权重，例如将 LinearLayout 里所有布局控件的 layout_weight 都设为 1，那么这些布局控件将平分该线性布局的宽度或者高度。

2. 为布局控件分配权重

在 LinearLayout 常用的属性中，使用 android:layout_weight 为布局控件分配权重是一个重点知识，接下来以图 2.2 所示的界面为例进行讲解。

例 2.1　LinearLayoutApp2_1

activity_main.xml 对应的代码如下所示。

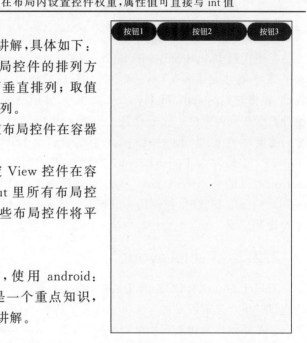

图 2.2　LinearLayout 的界面

```
<?xml version = "1.0" encoding = "utf - 8"?>
<LinearLayout xmlns:android = "http://schemas.android.com/apk/res/android"
    xmlns:tools = "http://schemas.android.com/tools"
```

```xml
    android:layout_width = "match_parent"
    android:layout_height = "match_parent"
    tools:context = ".MainActivity"
    android:orientation = "horizontal">

    <Button
        android:id = "@ + id/button"
        android:layout_width = "0dp"
        android:layout_height = "wrap_content"
        android:layout_weight = "1"
        android:text = "按钮 1" />

    <Button
        android:id = "@ + id/button2"
        android:layout_width = "0dp"
        android:layout_height = "wrap_content"
        android:layout_weight = "2"
        android:text = "按钮 2" />

    <Button
        android:id = "@ + id/button3"
        android:layout_width = "0dp"
        android:layout_height = "wrap_content"
        android:layout_weight = "1"
        android:text = "按钮 3" />
</LinearLayout>
```

上述代码中，android:orientation 属性值为 horizontal，表示在 LinearLayout 中的控件水平排列，这也是 LinearLayout 中控件默认的排列方式。同时在布局中定义了 3 个 Button 控件，并将它们的 android:layout_weight 属性值分别设置为 1、2、1，这就说明这 3 个 Button 控件占据布局的宽度占比分别是 1/4、2/4 和 1/4。

需要注意的是，LinearLayout 中的 android:layout_width 属性值不可设为 wrap_content。这是因为 LinearLayout 的优先级比 Button 高，如果设置为 wrap_content，则 Button 控件的 android:layout_weight 属性会失去作用。当设置了 Button 控件的 android:layout_weight 属性时，控件的 android:layout_width 属性值一般设置为 0dp 才会有权重占比的效果。

2.2.3 TableLayout

观看视频

TableLayout 将屏幕划分为表格，通过指定行和列可以将界面元素添加到表格中，表格的边界对用户是不可见的。TableLayout 的每一行是一个 TableRow 的对象，当然也可以是一个布局控件的对象。如果直接往 TableLayout 中添加控件，那么这个控件将占满一行。如果一行上有多个控件，就要添加一个 TableRow 容器，把控件都放到里面。

TableLayout 还支持嵌套，可以将另一个 TableLayout 放置在前一个 TableLayout 的表格中，也可以在 TableLayout 中添加其他界面布局，例如 LinearLayout、RelativeLayout 等。在 XML 布局文件中定义 TableLayout 的基本语法格式如下：

```
<TableLayout xmlns:android = "http://schemas.android.com/apk/res/android"
    属性 = "属性值">
    <TableRow>
        UI 控件
    </TableRow>
        UI 控件
        ……
</TableLayout>
```

TableLayout 继承自 LinearLayout,因此它完全支持 LinearLayout 所有的属性,此外,它还有其他常用属性,具体如表 2.3 所示。

表 2.3 TableLayout 的常用属性

属 性 名 称	功 能 描 述
android:collapseColumns	设置可被隐藏的列
android:shrinkColumns	设置可被收缩的列
android:stretchColumns	设置可被拉伸的列

TableLayout 常用属性说明如下:

(1) android:collapseColumns="1,2":设置需要被隐藏的列序号(序号从 0 开始)。列之间必须用逗号隔开,例如,1,2,5。

(2) android:shrinkColumns="1,2":设置允许被收缩的列的序号(序号从 0 开始)。

(3) android:stretchColumns="1,2":设置允许被拉伸的列的序号(序号从 0 开始),以填满剩下的多余空白空间,列之间必须用逗号隔开。

TableLayout 中的控件有两个常用属性,分别用于设置控件显示的位置、占据的行数,具体如表 2.4 所示。

表 2.4 TableLayout 中控件的常用属性

属 性 名 称	功 能 描 述
android:layout_column	设置该控件显示的位置
android:layout_span	设置该控件占据几列

对于<TableRow></TableRow>内的控件而言,下标从 0 开始,TableLayout 中控件常用属性说明如下:

(1) android:layout_column="1":表示该控件显示在第 2 列。

(2) android:layout_span="2":表示该控件占据 2 列。

接下来,以图 2.3 所示的界面为例进行讲解,帮助读者掌握 TableLayout 的用法。

例 2.2 TableLayoutApp2_2

activity_main.xml 对应的代码如下所示。

```
<?xml version = "1.0" encoding = "utf-8"?>
<TableLayout xmlns:android = "http://schemas.android.com/apk/res/android"
    xmlns:tools = "http://schemas.android.com/tools"
    android:layout_width = "match_parent"
    android:layout_height = "match_parent"
    tools:context = ".MainActivity"
    android:stretchColumns = "0">
```

图 2.3　TableLayout 的界面

```
<TableRow>
    <TextView
        android:text = "用户名"
        android:layout_width = "wrap_content"
        android:layout_height = "wrap_content"
        android:gravity = "right"
        android:layout_weight = "1"/>
    <EditText
        android:layout_width = "wrap_content"
        android:layout_height = "wrap_content"
        android:inputType = "textPersonName"
        android:layout_weight = "1"
        android:hint = "请输入用户名"/>
</TableRow>

<TableRow>
    <Button
        android:text = "确定"
        android:layout_width = "wrap_content"
        android:layout_height = "wrap_content"
        android:layout_weight = "1"/>
    <Button
        android:text = "确定"
```

```
            android:layout_width = "wrap_content"
            android:layout_height = "wrap_content"
            android:layout_weight = "1"/>
    </TableRow>
</TableLayout>
```

在上述代码中,共包含两个 TableRow:第一个 TableRow 里存放一个 TextView 控件和一个 EditText 控件;第二个 TableRow 里存放两个 Button 控件。另外,在布局文件中设置了 android:stretchColumns = "0",表示允许第 1 列被拉伸,以填满剩下的多余空白空间。

2.2.4 FrameLayout

FrameLayout 也叫框架布局,是一个非常简单的界面布局。这个布局就是直接在屏幕上开辟一块空白的区域,当向 FrameLayout 添加控件时默认放到左上角,FrameLayout 没有任何定位方式,不过可以为控件添加 layout_gravity 属性,从而指定组件的对齐方式。

在 FrameLayout 中,如果 FrameLayout 有多个控件,那么后放置的控件将遮挡先放置的控件,布局的大小则由子控件中最大的那个控件决定。如果所有控件都一样大,同一时刻就只能看到最上面的那个控件。可以使用 Android SDK 中提供的层级观察器(Hierarchy Viewer)进一步分析界面布局,层级观察器能够对用户界面进行分析和调试,并以图形化的方式展示树形结构的界面布局。在 XML 布局文件中定义 FrameLayout 的基本语法格式如下:

```
< FrameLayout xmlns:android = "http://schemas.android.com/apk/res/android"
     属性 = "属性值">
</FrameLayout>
```

FrameLayout 除了上面介绍的通用属性外,还有两个特殊的属性,具体如表 2.5 所示。

表 2.5　FrameLayout 属性

属 性 名 称	功 能 描 述
android:foreground	设置 FrameLayout 容器的前景图像(始终在所有子控件之上)
android:foregroundGravity	设置前景图像显示的位置

(1) android:foreground:设置 FrameLayout 容器的前景图片,其值为完整路径的图片。

(2) android:foregroundGravity:设置前景图片显示的位置,其值为 top、bottom、left、right、center_vertical、horizontal_vertical 等。

接下来,以图 2.4 所示的界面布局为例,讲解如何在布局中使用 android:foreground 和 android:foregroundGravity 属性指定控件位置。

例 2.3　FrameLayoutApp2_3

activity_main.xml 对应的代码如下所示。

```
<?xml version = "1.0" encoding = "utf - 8"?>
< FrameLayout xmlns:android = "http://schemas.android.com/apk/res/android"
    xmlns:tools = "http://schemas.android.com/tools"
```

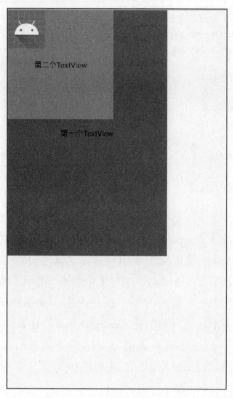

图 2.4 FrameLayout 的界面

```
        android:layout_width = "match_parent"
        android:layout_height = "match_parent"
        tools:context = ".MainActivity"
        android:foreground = "@mipmap/ic_launcher"
        android:foregroundGravity = "left">

    < TextView
            android:layout_width = "300dp"
            android:layout_height = "450dp"
            android:background = "#FF6143"
            android:gravity = "center"
            android:text = "第一个 TextView" />

    < TextView
            android:layout_width = "200dp"
            android:layout_height = "200dp"
            android:background = "#7BFE00"
            android:gravity = "center"
            android:text = "第二个 TextView" />
</FrameLayout>
```

在上述代码中,通过 android:foreground 和 android:foregroundGravity 属性设置 ic_launcher.png 为 FrameLayout 的前景图片并居左显示。前景图片始终保存在该布局最上层。在布局中定义了两个 TextView 控件,设置了不同的颜色,第二个 TextView 控件位于第一个 TextView 控件的上一层。

2.2.5 RelativeLayout

RelativeLayout 利用控件之间的相对位置关系来进行布局,相对位置关系主要是控件与父容器、控件与其他控件之间的相对关系。在 XML 布局文件中定义 RelativeLayout 的基本语法格式如下:

观看视频

```
<RelativeLayout xmlns:android = "http://schemas.android.com/apk/res/android"
    属性 = "属性值"
    ......>
</RelativeLayout>
```

RelativeLayout 的属性有很多,具体如表 2.6 所示。

表 2.6 RelativeLayout 的属性

属 性 名 称	功 能 描 述
android:layout_alignParentTop	将该控件的顶部与其父控件的顶部对齐
android:layout_alignParentBottom	将该控件的底部与其父控件的底部对齐
android:layout_alignParentLeft	将该控件的左部与其父控件的左部对齐
android:layout_alignParentRight	将该控件的右部与其父控件的右部对齐
android:layout_centerHorizontal	将该控件置于水平居中
android:layout_centerVertical	将该控件置于垂直居中
android:layout_centerInParent	将该控件置于父控件的中央
android:layout_above	将该控件的底部置于给定 ID 的控件之上
android:layout_below	将该控件的底部置于给定 ID 的控件之下
android:layout_toLeftOf	将该控件的右边缘与给定 ID 的控件左边缘对齐
android:layout_toRightOf	将该控件的左边缘与给定 ID 的控件右边缘对齐
android:layout_alignBaseline	将该控件的基准线与给定 ID 的基准线对齐
android:layout_alignTop	将该控件的顶部边缘与给定 ID 的顶部边缘对齐
android:layout_alignBottom	将该控件的底部边缘与给定 ID 的底部边缘对齐
android:layout_alignLeft	将该控件的左边缘与给定 ID 的左边缘对齐
android:layout_alignRight	将该控件的右边缘与给定 ID 的右边缘对齐

接下来,以图 2.5 所示的界面为例,讲解如何在 RelativeLayout 中使用这些属性。

例 2.4 RelativeLayoutApp2_4

activity_main.xml 对应的代码如下所示。

```
<?xml version = "1.0" encoding = "utf-8"?>
<RelativeLayout xmlns:android = "http://schemas.android.com/apk/res/android"
    xmlns:tools = "http://schemas.android.com/tools"
    android:layout_width = "match_parent"
    android:layout_height = "match_parent"
    tools:context = ".MainActivity">

    <TextView
        android:id = "@ + id/center"
        android:layout_width = "100dp"
        android:layout_height = "100dp"
```

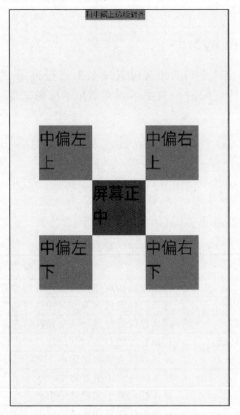

图 2.5 RelativeLayout 的界面

```
        android:layout_centerInParent = "true"
        android:background = "#ff0000"
        android:text = "屏幕正中"
        android:textSize = "30sp" />

    <TextView
        android:layout_width = "100dp"
        android:layout_height = "100dp"
        android:layout_above = "@id/center"
        android:layout_toLeftOf = "@id/center"
        android:background = "#00ff00"
        android:text = "中偏左上"
        android:textSize = "30sp" />

    <TextView
        android:layout_width = "100dp"
        android:layout_height = "100dp"
        android:layout_above = "@id/center"
        android:layout_toRightOf = "@id/center"
        android:background = "#00ff00"
        android:text = "中偏右上"
        android:textSize = "30sp" />

    <TextView
```

```
            android:layout_width = "100dp"
            android:layout_height = "100dp"
            android:layout_below = "@id/center"
            android:layout_toLeftOf = "@id/center"
            android:background = "#00ff00"
            android:text = "中偏左下"
            android:textSize = "30sp" />

    <TextView
            android:layout_width = "100dp"
            android:layout_height = "100dp"
            android:layout_below = "@id/center"
            android:layout_toRightOf = "@id/center"
            android:background = "#00ff00"
            android:text = "中偏右下"
            android:textSize = "30sp" />

    <TextView
            android:layout_width = "wrap_content"
            android:layout_height = "wrap_content"
            android:layout_alignRight = "@id/center"
            android:background = "#00ff00"
            android:text = "和中间上边线对齐" />
</RelativeLayout>
```

在上述代码中,定义了 RelativeLayout,通过设置 android:layout_width 和 android:layout_height 属性的值设置该布局的宽高。首先定义了 TextView 控件,通过设置 android:layout_centerInParent="true"将控件设置在布局的中央,并设置了红色的背景颜色。

接着又定义了三个不同背景颜色的 TextView 控件,通过控件的属性来设置这三个控件在中央控件中的不同位置,具体如下。

(1) android:layout_above="@id/center"、android:layout_toLeftOf="@id/center": 通过这两个属性设置控件位于中央控件的左上方。

(2) android:layout_above="@id/center"、android:layout_toRightOf="@id/center":通过这两个属性设置控件位于中央控件的右上方。

(3) android:layout_below="@id/center"、android:layout_toLeftOf="@id/center": 通过这两个属性设置控件位于中央控件的左下方。

(4) android:layout_below="@id/center"、android:layout_toRightOf="@id/center":通过这两个属性设置控件位于中央控件的右下方。

最后定义一个 TextView 控件,通过属性 android:layout_alignRight="@id/center"将控件设置在布局中间与上边线对齐的位置。

2.2.6 GridLayout

GridLayout 类似于表格布局(TableLayout),是 Android 4.0 及以上版本新增加的布局,主要以网格的形式来布局窗口控件。使用虚细线将布局划分为行、列和单元格,也支持一个控件在行、列上交错排列,分为水平和垂直两种方式,默认是水平布局,一个控件挨着一个控件从左到右依次排列。在 XML 布局文件中定义 GridLayout 的基本语法格式如下：

```
<GridLayout xmlns:android = "http://schemas.android.com/apk/res/android"
    属性 = "属性值"
    ……>
</GridLayout>
```

GridLayout 属性如表 2.7 所示。

表 2.7 GridLayout 属性

属 性 名 称	功 能 描 述
rowCount	设置 GridLayout 有多少行
columnCount	设置 GridLayout 有多少列
layout_row	子控件在布局的行数
layout_column	子控件在布局的列数
layout_rowSpan	跨行数
layout_columnSpan	跨列数

接下来，以图 2.6 所示的界面布局为例，讲解 GridLayout 属性的设置。

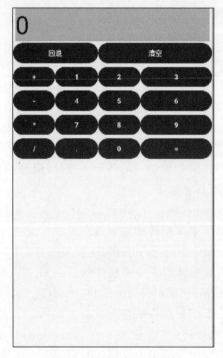

图 2.6 GridLayout 的界面

例 2.5 GridLayoutApp2_5

activity_main.xml 对应的代码如下所示。

```
<?xml version = "1.0" encoding = "utf-8"?>
<GridLayout xmlns:android = "http://schemas.android.com/apk/res/android"
    xmlns:tools = "http://schemas.android.com/tools"
    android:layout_width = "fill_parent"
```

```xml
        android:layout_height = "wrap_content"
        android:columnCount = "4"
        android:orientation = "horizontal"
        android:rowCount = "6"
        tools:context = ".MainActivity">

    < TextView
            android:layout_columnSpan = "4"
            android:layout_gravity = "fill"
            android:layout_marginLeft = "5dp"
            android:layout_marginRight = "5dp"
            android:background = "#FFCCCC"
            android:text = "0"
            android:textSize = "50sp" />

    < Button
            android:layout_columnSpan = "2"
            android:layout_gravity = "fill"
            android:text = "回退"
            android:textStyle = "bold" />

    < Button
            android:layout_columnSpan = "2"
            android:layout_gravity = "fill"
            android:text = "清空"
            android:textStyle = "bold"/>

    < Button
            android:layout_gravity = "fill"
            android:text = " + "
            android:textStyle = "bold"/>

    < Button
            android:layout_gravity = "fill"
            android:text = "1"
            android:textStyle = "bold"/>

    < Button
            android:layout_gravity = "fill"
            android:text = "2"
            android:textStyle = "bold"/>

    < Button
            android:layout_gravity = "fill"
            android:text = "3"
            android:textStyle = "bold"/>

    < Button
            android:layout_gravity = "fill"
            android:text = " - "
            android:textStyle = "bold"/>

    < Button
            android:layout_gravity = "fill"
```

```xml
            android:text = "4" />

    < Button
        android:layout_gravity = "fill"
        android:text = "5"
        android:textStyle = "bold"/>

    < Button
        android:layout_gravity = "fill"
        android:text = "6"
        android:textStyle = "bold"/>

    < Button
        android:layout_gravity = "fill"
        android:text = " * "
        android:textStyle = "bold"/>

    < Button
        android:layout_gravity = "fill"
        android:text = "7"
        android:textStyle = "bold"/>

    < Button
        android:layout_gravity = "fill"
        android:text = "8"
        android:textStyle = "bold"/>

    < Button
        android:layout_gravity = "fill"
        android:text = "9"
        android:textStyle = "bold"/>

    < Button
        android:layout_gravity = "fill"
        android:text = "/"
        android:textStyle = "bold"/>

    < Button
        android:layout_width = "wrap_content"
        android:layout_gravity = "fill"
        android:text = "."
        android:textStyle = "bold"/>

    < Button
        android:layout_gravity = "fill"
        android:text = "0"
        android:textStyle = "bold"/>

    < Button
        android:layout_gravity = "fill"
        android:text = " = "
        android:textStyle = "bold"/>
</GridLayout >
```

在上述代码中实现了一个简易计算器的界面，首先使用了一个 TextView 控件模拟计算显示数据，并将其设置为占据 4 列(layout_columnSpan="4")。

接着定义了一个 4 列(columnCount="4")6 行(rowCount="6")的水平 GridLayout，并将其宽度设置为填满屏幕(fill_parent)，高度设置为包裹内容。

最后添加计算器所需的 Button 控件，"回退"与"清空"按钮横跨两列，使用 GridLayout 子控件的属性 layout_columnSpan="2"进行设置，其他按钮都是直接添加，占用一行一列。需要注意的是，通过 android:layout_rowSpan 与 android:layout_columnSpan 设置了组件横跨多行或者多列时，如果想要让组件填满横越过的行或列，则可以设置 android:layout_gravity = "fill"。

2.2.7 AbsoluteLayout

AbsoluteLayout 能通过指定界面元素的坐标位置，来确定用户界面的整体布局，又称为坐标布局。在 XML 布局文件中定义 AbsoluteLayout 的基本语法格式如下：

```
< AbsoluteLayout xmlns:android = "http://schemas.android.com/apk/res/android">
    ......
</AbsoluteLayout >
```

AbsoluteLayout 常用的属性主要有两个，具体如表 2.8 所示。

表 2.8　AbsoluteLayout 常用的属性

属性名称	功能描述
layout_x	控件在坐标系 x 轴的位置
layout_y	控件在坐标系 y 轴的位置

在 Android 手机中，坐标系以手机屏幕左上角的顶点为坐标原点，从该点向右为 x 轴的正方向，从该点向下则为 y 轴的正方向。

需要说明的是，由于 Android 设备的分辨率种类繁多，小到 320×240 像素，大到 1920×1080 像素等，差异较大，而 AbsoluteLayout 通过 x 轴和 y 轴确定界面元素位置后，Android 系统不能够根据不同屏幕对界面元素的位置进行调整，从而降低了界面布局对不同类型和尺寸屏幕的适应能力。因此，现在 AbsoluteLayout 是一种不推荐使用的界面布局。但是从学习知识的角度，我们还是有必要来了解这种布局方式，接下来以图 2.7 所示的界面布局为例，讲解 AbsoluteLayout 的用法。

例 2.6　AbsoluteLayoutApp2_6

activity_main.xml 对应的代码如下所示。

```
<?xml version = "1.0" encoding = "utf - 8"?>
< AbsoluteLayout xmlns:android = "http://schemas.android.com/apk/res/android"
    xmlns:tools = "http://schemas.android.com/tools"
    android:layout_width = "match_parent"
    android:layout_height = "match_parent"
    android:orientation = "vertical"
    tools:context = ".MainActivity">
```

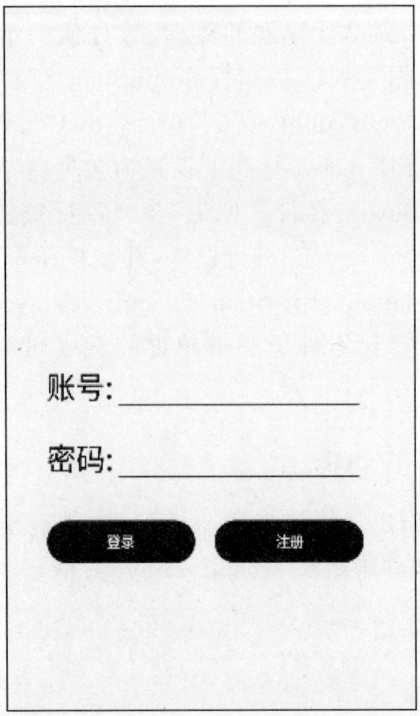

图 2.7 AbsoluteLayout 的界面

```
< TextView
      android:layout_width = "wrap_content"
      android:layout_height = "wrap_content"
      android:layout_x = "40dp"
      android:layout_y = "350dp"
      android:text = "账号："
      android:textSize = "30sp" />

< EditText
      android:layout_width = "250dp"
      android:layout_height = "50dp"
      android:layout_x = "110dp"
      android:layout_y = "350dp" />

< TextView
      android:layout_width = "wrap_content"
      android:layout_height = "wrap_content"
      android:layout_x = "40dp"
      android:layout_y = "420dp"
      android:text = "密码："
      android:textSize = "30sp" />

< EditText
      android:layout_width = "250dp"
      android:layout_height = "50dp"
      android:layout_x = "110dp"
      android:layout_y = "420dp" />

< Button
      android:layout_width = "150dp"
      android:layout_height = "50dp"
```

```
        android:layout_x = "40dp"
        android:layout_y = "500dp"
        android:text = "登录"
        android:textStyle = "bold" />

    < Button
        android:layout_width = "150dp"
        android:layout_height = "50dp"
        android:layout_x = "210dp"
        android:layout_y = "500dp"
        android:text = "注册"
        android:textStyle = "bold" />

</AbsoluteLayout >
```

在上述代码中,使用 AbsoluteLayout 完成了一个简易的登录界面,在布局中添加了 TextView、EditText、Button 控件,主要使用坐标来固定控件显示的位置。

2.2.8 ConstraintLayout

ConstraintLayout 是 Google 年在 2016 的 I/O 大会上推出的全新的布局,是 Android Studio 2.3+ 创建 activity_main.xml 的默认布局方式。它根据布局中其他控件或视图,确定控件在屏幕中的位置,受到其他控件、父容器和基准线三类约束,简单理解认为它是 RelativeLayout 的超级加强升级版本。它可以有效解决布局嵌套的性能问题和复杂布局的快速实现,性能大约提高 40%,另外实现布局的方法也十分简单,通过 Android Studio 提供的编辑器,简单拖曳就可以快速实现布局。

1. 相对定位

相对定位是在 ConstraintLayout 中创建布局的基本约束之一。这些约束允许用户将指定的控件(源控件)相对于另一个控件(目标控件)进行定位。用户可以在水平和垂直轴上约束控件,其中水平轴包括 left、start、right、end,垂直轴包括 top、bottom、baseline(文本底部的基准线)。例如将控件 B 约束到控件 A 的右侧,如图 2.8 所示。

图 2.8 中,将控件 B 左侧的边约束到控件 A 右侧的边,从而将控件 B 定位到控件 A 的右侧。这里的约束可以理解为边的对齐。

在图 2.9 中,每一条边(top、bottom、baseline、left、start、right、end)都可以与其他控件形成约束,这些边形成相对定位关系的属性如表 2.9 所示。

图 2.8 相对定位的约束

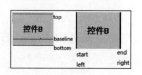

图 2.9 控件的约束边

表 2.9 相对定位关系的属性

属 性 名 称	功 能 描 述
layout_constraintLeft_toLeftOf	控件的左边与另一个控件的左边对齐
layout_constraintLeft_toRightOf	控件的左边与另一个控件的右边对齐

续表

属 性 名 称	功 能 描 述
layout_constraintRight_toLeftOf	控件的右边与另一个控件的左边对齐
layout_constraintRight_toRightOf	控件的右边与另一个控件的右边对齐
layout_constraintTop_toTopOf	控件的上边与另一个控件的上边对齐
layout_constraintTop_toBottomOf	控件的上边与另一个控件的下边对齐
layout_constraintBottom_toTopOf	控件的下边与另一个控件的上边对齐
layout_constraintBottom_toBottomOf	控件的下边与另一个控件的下边对齐
layout_constraintBaseline_toBaselineOf	控件间的文本内容基准线对齐
layout_constraintStart_toEndOf	控件的起始边与另一个控件的尾部对齐
layout_constraintStart_toStartOf	控件的起始边与另一个控件的起始边对齐
layout_constraintEnd_toStartOf	控件的尾部与另一个控件的起始边对齐
layout_constraintEnd_toEndOf	控件的尾部与另一个控件的尾部对齐

2. 居中定位和偏移

在 ConstraintLayout 中，通过添加约束，不仅可以确定两个控件的相对位置，也可以确定某控件在父布局中的相对位置。若相同方向上（横轴或纵轴）控件两边同时向 ConstraintLayout 添加约束，则控件在添加约束的方向上居中显示。在父布局中横向居中显示的控件如图 2.10 所示。

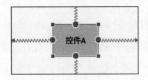

图 2.10　控件居中显示

在约束是同向相反的情况下，默认控件是居中的，但是也像拔河一样，两个约束的力大小不等时，就会产生偏移，偏移的属性如表 2.10 所示。

表 2.10　偏移的属性

属 性 名 称	功 能 描 述
layout_constraintHorizontal_bias	横轴方向偏移
layout_constraintVertical_bias	纵轴方向偏移

3. 链

链（Chain）是一种特殊的约束，它使我们能够对一组水平或竖直方向互相关联的控件进行统一管理。一组控件通过一个双向的约束关系链接起来，就能形成一个链。形成的链如图 2.11 所示。

在图 2.10 中，链中的第一个控件 A 称为头控件，链的头控件可以通过 layout_constraintHorizontal_chainStyle 和 layout_constraintVertical_chainStyle 属性设置水平链条和竖直链条的样式。其属性值为 spread、spread_inside 和 packed，具体如下：

（1）spread：设置控件在布局内平均分布。其为链的默认样式，如图 2.11 所示。

（2）spread_inside：设置链的两段贴近父容器，其他的控件将在剩余的空间内采用 spread 样式进行布局，如图 2.12 所示。

（3）packed：设置链中的所有控件合并在一起后在布局内居中显示，如图 2.13 所示。

图 2.11　链约束　　　　图 2.12　spread_inside 样式　　　　图 2.13　packed 样式

ConstraintLayout 中,当控件宽或者高的属性设置为 0dp 时,链的 3 种样式可以搭配 layout_constraintHorizontal_weight 属性形成权重链(Weighted Chain)的样式。packed 可以搭配 layout_constraint Horizontal_bias 属性控制链与父容器的间距从而形成 Packed Chain With Bias(带偏置的填充链)样式。

2.3 小结

- Android 应用的界面是由布局和控件构成的,可以使用 XML 文件控制 UI 界面,也可以使用 Java 代码来控制,第一种在开发中比较常用。
- 线性布局(LinearLayout)主要将控件以水平或者垂直的方式进行排列显示,如果对水平排列的控件设置权重,一般将其宽度设置为 0dp。
- 相对布局(RelativeLayout)主要以父容器或者其他控件为参照物,来设置控件在布局中的显示位置,这样可以更加方便地设置各控件之间的相对位置、距离等。
- 帧布局(FrameLayout)类似于 Photoshop 中图层的概念,为每个加入的控件都创建单独的帧,看上去像是控件叠加在一起。
- 网格布局(GridLayout)使用虚细线将布局划分为行、列和单元格,也支持一个控件在行、列上交错排列。可以自己设置布局中组件的排列方式,自定义网格布局有多少行、多少列,直接设置组件位于某行某列,设置组件横跨几行或者几列。
- 表格布局(TableLayout)以行和列的形式对控件进行管理,每一行为一个 TableRow 对象,或一个 View 控件。当为 TableRow 对象时,可在 TableRow 下添加子控件,默认情况下,每个子控件占据一列;当为 View 控件时,该 View 将独占一行。
- 绝对布局(AbsoluteLayout)直接指定控件的坐标,由于不能适应屏幕大小的变化,一般不推荐使用。
- 约束布局(ConstraintLayout)的出现主要是为了解决布局嵌套过多的问题,更加倾向于可视化操作。

习题 2

1. 下列属性中,用于设置线性布局方向的是(   )。
   A. orientation   B. gravity   C. layout_gravity   D. padding
2. 下列选项中,不属于 Android 布局的是(   )。
   A. FrameLayout   B. LinearLayout   C. Button   D. RelativeLayout
3. 对于 XML 布局文件,android:layout_width 属性的值不可以是(   )。
   A. match_parent   B. fill_parent   C. warp_content   D. match_content
4. 下列关于 RelativeLayout 的描述,正确的是(   )。
   A. RelativeLayout 表示绝对布局,可以自定义控件的 x、y 位置
   B. RelativeLayout 表示帧布局,可以实现标签切换的功能
   C. RelativeLayout 表示相对布局,其中控件的位置都是相对位置
   D. RelativeLayout 表示表格布局,需要配合 TableRow 一起使用

5. Android 的常见布局都直接或者间接地继承自_____类。
6. 表格布局(TableLayout)通过_____布局拉制表格的行数。
7. _____布局通过相对定位的方式指定子控件的位置。
8. 在 R.java 文件中,android:id 属性会自动生成对应的_____类型的值。
9. 列举 Android 中的常用布局,并简述它们各自的特点。
10. 按照本章中 GridLayout 实现的计算器界面,用其他多种方式进行实现。

第3章 Android常见界面控件

本章导图

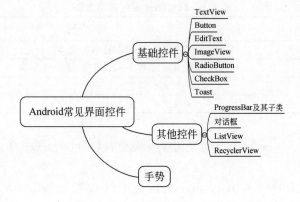

主要内容

- 基础控件的使用。
- ProgressBar 及其子类。
- 对话框的使用。
- ListView 的使用。
- RecyclerView 的使用。
- 手势的使用。

难点

- ListView 的使用。
- 数据适配器的使用方法。
- RecyclerView 的使用。

Android 系统提供了丰富的 UI 组件用于程序设计。在 Android Studio 中一般可以通过拖曳的方式对组件进行布局。各种组件都有一系列的属性和方法,通过这些属性和方法可以方便地操纵组件。对于具有事件触发的组件而言,开发人员可以设置事件监听器进行响应。本章将针对 Android 常见的界面控件进行讲解。

3.1 基础控件的使用

用户界面是系统与用户之间进行信息交互的接口，Android 借用了 Java 中的界面设计思想及事件响应机制。Android 系统为程序员提供了丰富的用户界面组件，包括菜单、对话框、按钮、文本框、下拉列表等。Android 支持控件拖放、XML 源码设计和程序代码操作 3 种设计形式。

3.1.1 TextView

TextView 控件用于显示文本信息。其常用方法如表 3.1 所示。

表 3.1 TextView 控件常用方法

方法	说明
getText()	用于获取控件中显示的文本
setText(text)	将 text 设置为控件中要显示的文本
setTextColor()	设置文本颜色
setTextSize()	设置文本字体大小

1．创建程序

创建一个名为 TextViewAPP3_1 的应用程序，指定包名为 com.example.TextViewAPP3_1。

2．放置界面控件

在 res/layout 文件夹的 activity_main.xml 文件中放置一个 TextView 控件，用于显示文本信息。activitymain.xml 文件的具体代码如【文件 3_1】所示。

【文件 3_1】

```xml
<?xml version = "1.0" encoding = "utf-8"?>
<RelativeLayout xmlns:android = "http://schemas.android.com/apk/res/android"
    xmlns:app = "http://schemas.android.com/apk/res-auto"
    xmlns:tools = "http://schemas.android.com/tools"
    android:layout_width = "match_parent"
    android:layout_height = "match_parent"
    tools:context = ".MainActivity">

    <TextView
        android:id = "@+id/textView"
        android:layout_width = "wrap_content"
        android:layout_height = "wrap_content"
        android:layout_centerInParent = "true"
        android:text = "Hello everyone!"
        android:textSize = "34sp" />
</RelativeLayout>
```

3.1.2 Button

Button(按钮)控件主要用于响应用户点击并引发点击事件。Button 控件表示按钮，它继承自 TextView 控件，既可以显示文本，又可以显示图片，同时也允许用户通过点击来执行操作。当 Button 控件被点击时，被按下与弹起的背景会有一个动态的切换效果，这个效果就是点击效果。

通常情况下，所有控件都可以设置点击事件，Button 控件也不例外，Button 控件最重要的作用就是响应用户的一系列点击事件。

为 Button 控件设置点击事件的方式主要有以下三种。

（1）在布局文件中指定 onClick 属性。可以在布局文件中指定 onClick 属性的值来设置 Button 控件的点击事件，示例代码如下。

```
< Button
    android:id = "@ + id/button"
    android:layout_width = "wrap_content"
    android:layout_height = "wrap_content"
    android:layout_gravity = "center_horizontal"
    android:onClick = "click"
    android:text = "方式一" />
```

上述代码中，Button 控件指定了 onClick 属性，可以在 Activity 中定义专门的方法来实现 Button 控件的点击事件。需要注意的是，在 Activity 中定义实现点击事件的方法名必须与 onClick 属性的值保持一致。

（2）使用匿名内部类。在 Activity 中，可以使用匿名内部类为 Button 控件设置点击事件，示例代码如下。

```
button2.setOnClickListener(new View.OnClickListener() {
    @Override
    public void onClick(View view) {
        button2.setText("您选择了方式二点击按钮");
    }
});
```

上述代码中，通过 Button 控件设置 setOnClickListener()方法实现对 Button 控件点击事件的监听。SetOnClickListener()方法中传递的参数是一个匿名内部类。如果监听到按钮被点击，那么就会调用匿名内部类中的 onClick()方法实现对 Button 控件的点击事件。

（3）用 Activity 实现 OnClickListener 接口。用当前 Activity 实现 View.OnClickListener 接口，同样可以为 Button 控件设置点击事件，示例代码如下。

```
public class MainActivity extends AppCompatActivity implements View.OnClickListener{

    private Button button3;

    protected void onCreate(Bundle savedInstanceState) {
        super.onCreate(savedInstanceState);
        setContentView(R.layout.activity_main);

        button3 = findViewById(R.id.button3);
        button3.setOnClickListener(this);
    }
```

```
    @Override
        public void onClick(View v){
            button3.setText("您选择了方式三点击按钮");
        }
}
```

按钮被点击之前如图 3.1 所示,三个按钮都被点击之后的运行结果如图 3.2 所示。

图 3.1 运行前

图 3.2 运行结果

3.1.3 EditText

EditText(编辑框)控件可用于输入、显示、编辑字符串。其常用方法如表 3.2 所示。

表 3.2 EditText 控件常用方法

方法	说明
getText()	用于获取控件中显示的文本
setText(text)	将 text 设置为控件中要显示的文本

方　　法	说　　明
setTextColor()	设置文本颜色
setHintTextColor()	设置提示信息文本颜色

以下是编辑框的 XML 标签定义示例：

```xml
<EditText
    android:id = "@+id/editTextText"
    android:layout_width = "wrap_content"
    android:layout_height = "wrap_content"
    android:layout_below = "@id/textView"
    android:layout_centerHorizontal = "true"
    android:ems = "10"
    android:inputType = "text"
    android:text = "Name" />
```

3.1.4 ImageView

ImageView(图像视图)控件用于显示图片信息。其常用方法如表 3.3 所示。

表 3.3　ImageView 控件常用方法

方　　法	说　　明
setImageURI(Uri uri)	设置 ImageView 所显示内容为指定 Uri
setMaxHeight(int h)	设置控件最大高度
setMaxWeight(int w)	设置控件最大宽度

以下是 ImageView 控件的 XML 标签定义示例：

```xml
<ImageView
    android:id = "@+id/imageView"
    android:layout_width = "wrap_content"
    android:layout_height = "wrap_content"
    android:layout_above = "@id/textView"
    android:layout_centerHorizontal = "true"
    tools:srcCompat = "@tools:sample/avatars" />
```

3.1.5 RadioButton

RadioButton(单选按钮)控件一般以按钮组的形式存在，只能在给定的系列选项组中选中一项，在设计时使用 RadioGroup(单选按钮组)控件对其进行分组。用户选中某个选项时，控件也将产生点击事件。如果单选按钮控件以按钮组的形式存在，单选按钮组控件将产生 onUncheckedChangeListener 事件。

以下是 RadioButton 和 RadioGroup 控件的 XML 标签定义示例：

```xml
<RadioGroup
    android:id = "@+id/radiogroup"
    android:layout_width = "match_parent"
    android:layout_height = "wrap_content"
```

观看视频

```
        android:orientation = "vertical">
    < RadioButton
        android:id = "@ + id/rbutton1"
        android:layout_width = "wrap_content"
        android:layout_height = "wrap_content"
        android:textSize = "30dp"
        android:text = "对"/>
    < RadioButton
        android:id = "@ + id/rbutton2"
        android:layout_width = "wrap_content"
        android:layout_height = "wrap_content"
        android:textSize = "30dp"
        android:text = "错"/>
</RadioGroup>
```

RadioGroup 控件常常通过 OnCheckedChangeListener() 方法来响应按钮组中的某个按钮被选中,使用方法如下所示,图 3.3 是按钮被选中后的运行结果。

```
radioGroup.setOnCheckedChangeListener(new RadioGroup.
OnCheckedChangeListener() {
        int checkedId) {
    //判断点击的是哪个 RadioButton
    if (checkedId == R.id.rbutton1) {
        textView.setText("您的回答是:对");
    } else {
        textView.setText("您的回答是:错");
    }
    }
});
```

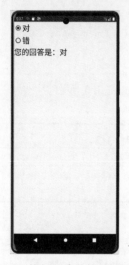

图 3.3 按钮组中的按钮被选中后的结果

3.1.6 CheckBox

CheckBox(复选框)控件是可以在给定的一系列选项中选中多项的控件。其常用方法如表 3.4 所示。

表 3.4 CheckBox 控件常用方法

方 法	说 明
isChecked()	判断复选框是否被选中
setChecked()	设置复选框状态
setOnClickListener()	设置复选框的点击事件监听器
setOnCheckedChangeListener()	设置复选框状态改变监听器
toggle()	改变复选框当前选中的状态

CheckBox 的 XML 标签定义示例如下所示:

```
< CheckBox
    android:id = "@ + id/like_red"
    android:layout_width = "wrap_content"
    android:layout_height = "wrap_content"
```

```
        android:text = "红"
        android:textSize = "18sp"/>
< CheckBox
        android:id = "@ + id/like_yellow"
        android:layout_width = "wrap_content"
        android:layout_height = "wrap_content"
        android:text = "黄"
        android:textSize = "18sp"/>
```

CheckBox 控件的复选事件是由 OnCheckedChangeListener()方法来监听并做出响应的,运行结果如图 3.4 所示。

3.1.7 Toast

观看视频

Toast 比较适合向用户显示系统运行中的状态消息,这类消息的重要性级别一般比较低,不太需要用户过多关注,如通知用户下载已完成。消息的显示过程中,它不会将焦点从 Activity 上移开,经过一段时间后,Toast 提示框会自动消失。由于无法保证用户会完全注意 Toast 消息提示,因此关键信息不能使用 Toast 消息提示。

用户可以调用 Toast 类的方法进行 Toast 消息提示,以下是调用 makeText()方法来显示短消息的,运行结果如图 3.5 所示。

图 3.4　CheckBox 控件运行结果　　　　图 3.5　Toast 控件运行结果

```
Toast.makeText(MainActivity.this," 电量不足 20 %,请充电",
Toast.LENGTH_SHORT).show();
```

3.2 ProgressBar 及其子类

ProgressBar(进度条)控件用于指示操作进度的用户界面元素,它以非中断方式向用户显示进度条。一般在应用的用户界面或通知中显示进度条,而不是在对话框中显示。

3.2.1 ProgressBar 的功能和用法

ProgressBar 支持两种表示进度的模式:不确定和确定。当不确定操作需要多长时间时,使用进度条的不确定模式。不确定模式是进度条的默认模式,并显示没有指定特定进度的循环动画。

如果要显示总量和已完成量确定的进度,则使用进度条的确定模式。例如,正在检索的文件的剩余百分比,写入数据库的批处理中的记录数量,或正在播放的音频文件的剩余百分比。其常用方法如表 3.5 所示。

表 3.5 ProgressBar 控件常用方法

方 法	说 明
setProgress(int value)	更新当前的进度
setMax(int max)	设置最大进度值
setIndeterminate(boolean indeterminate)	设置进度的模式:true 为不确定模式;false 为确定模式

以下演示了一个通过按下按钮触发进度条开始更新进度的例子,为了模拟进度条滚动的效果,启动了新线程,并在其中实现了进度的更新。以下是布局文件。

```xml
<?xml version = "1.0" encoding = "utf-8"?>
<LinearLayout xmlns:android = "http://schemas.android.com/apk/res/android"
    xmlns:app = "http://schemas.android.com/apk/res-auto"
    xmlns:tools = "http://schemas.android.com/tools"
    android:layout_width = "match_parent"
    android:layout_height = "match_parent"
    android:orientation = "vertical"
    tools:context = ".MainActivity" >

    <Button
        android:id = "@+id/button"
        android:layout_width = "wrap_content"
        android:layout_height = "wrap_content"
        android:layout_gravity = "center"
        android:onClick = "startProgram"
        android:text = "开始" />

    <ProgressBar
        android:id = "@+id/progressBar"
        style = "?android:attr/progressBarStyleHorizontal"
        android:layout_width = "match_parent"
        android:layout_height = "wrap_content"
        android:progress = "5"/>
</LinearLayout>
```

以下是实现的代码。

```java
public class MainActivity extends AppCompatActivity {

    @Override
    protected void onCreate(Bundle savedInstanceState) {
        super.onCreate(savedInstanceState);
        setContentView(R.layout.activity_main);
    }

    public void startProgram(View view){
        final ProgressBar progressBar = (ProgressBar)findViewById(R.id.progressBar);
        final Thread t = new Thread(){
            @Override
            public void run() {
                super.run();
                int i = 0;
                try {
                    while(i <= 100){
                        progressBar.setProgress(i);
                        i += 5;
                        sleep(1000);
                    }
                }catch (Exception e){
                    Log.e("ProgressBar", e.toString());
                }
            }
        };
        t.start();
    }
}
```

图 3.6 展示了以上例子的运行结果。

3.2.2　SeekBar 的功能和用法

SeekBar（拖动条）和进度条非常相似，只是进度条采用颜色填充来表明进度完成的程度，而拖动条则通过滑块的位置来标识数值——而且拖动条允许用户拖动滑块来改变值，因此拖动条通常用于对系统的某种数值进行调节，如调节音量等。

由于 SeekBar 继承了 ProgressBar，因此 ProgressBar 所支持的 XML 属性和方法完全适用于 SeekBar。

SeekBar 允许用户改变拖动条的滑块外观，改变滑块外观通过如下属性来指定。

- android:thumb：指定一个 Drawable 对象，该对象将作为自定义滑块。
- android:tickMark：指定一个 Drawable 对象，该对象将作为自定义刻度图标。

为了让程序能响应拖动条滑块位置的改变，可以考虑为它绑定一个 OnSeekBarChangeListener 监听器。

SeekBar 控件的 XML 标签定义示例如下所示：

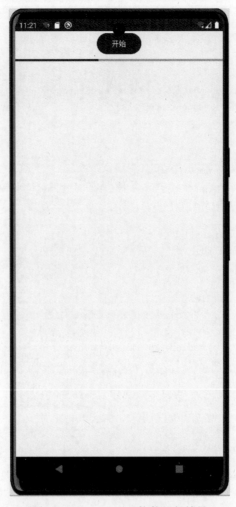

图 3.6　ProgressBar 控件运行结果

```
< SeekBar
    android:id = "@ + id/seekBar"
    android:layout_width = "match_parent"
    android:layout_height = "wrap_content"
    android:max = "100"
    android:progress = "1"
    android:thumb = "@mipmap/ic_launcher" />
```

图 3.7 显示了 SeekBar 控件的运行结果。

3.2.3　RatingBar 的功能和用法

RatingBar(星级评分条)与 SeekBar 有相同的父类：AbsSeekBar,因此它们十分相似。它们都允许用户通过拖动来改变进度,最大的区别是 RatingBar 通过星星来表示进度。以下是 RatingBar 控件的 XML 标签定义示例,图 3.8 展示了 Ratingbar 控件的运行结果。

图 3.7　SeekBar 控件运行结果　　　　图 3.8　RatingBar 控件运行结果

```
<RatingBar
    android:id = "@ + id/ratingBar"
    android:layout_width = "wrap_content"
    android:layout_height = "wrap_content"
    android:numStars = "8"
    android:stepSize = "0.5"
    android:progress = "1"
    android:max = "8"/>
```

3.3　对话框的使用

在 Android 应用程序中，AlertDialog 对话框用于提示一些重要信息或者显示一些需要用户额外交互的内容。它一般以小窗口的形式展示在界面上。

3.3.1 使用 AlertDialog 建立对话框

使用 AlertDialog 创建的对话框一般包含标题、内容和按钮三个区域。一般情况下，创建 AlertDialog 对话框的步骤大致分为以下几步。

（1）调用 AlertDialog 静态内部类 Builder，创建 AlertDialog.Builder 对象。

（2）调用 AlertDialog.Builder 对象的 setTitle()和 setIcon()方法分别设置 AlertDialog 对话框的标题名称和图标。

（3）调用 AlertDialog.Builder 对象的 setMessage()、setSingleChoiceItems()方法或 setMultiChoiceItems()方法设置 AlertDialog 对话框的内容为简答文本、单选列表或者多选列表。

（4）调用 AlertDialog.Builder 对象的 setPositiveButton()和 setNegativeButton()方法设置 AlertDialog 对话框的"确定"和"取消"按钮。

（5）调用 AlertDialog.Builder 对象的 create()方法创建 AlertDialog 对象。

（6）调用 AlertDialog 对象的 show()方法显示该对话框。

（7）调用 AlertDialog 对象的 dismiss()方法取消该对话框。

以下代码通过点击手机上的后退键弹出一个对话框，运行结果如图 3.9 所示。

```java
@Override
public void onBackPressed() {
    //声明对象
    AlertDialog dialog;
    AlertDialog.Builder builder = new AlertDialog.Builder(this)
            .setTitle("普通对话框")              //设置对话框的标题
            .setIcon(R.mipmap.ic_launcher)     //设置标题图标
            .setMessage("是否确定退出应用?")    //设置对话框的提示信息
            //添加"确定"按钮
            .setPositiveButton("确定", new DialogInterface.OnClickListener() {
                @Override
                public void onClick(DialogInterface dialog, int which) {
                    dialog.dismiss();           //关闭对话框
                    MainActivity.this.finish(); //关闭 MainActivity
                }
            })
            //添加"取消"按钮
            .setNegativeButton("取消", new DialogInterface.OnClickListener() {
                @Override
                public void onClick(DialogInterface dialog, int which) {
                    dialog.dismiss();
                }
            });
    dialog = builder.create();
    dialog.show();
}
```

3.3.2 创建单选和多选对话框

单选对话框的内容区域显示为单选列表。只需要调用 AlertDialog.Builder 对象的

图 3.9 普通对话框运行结果

setSingleChoiceItems()方法即可创建带单选列表的对话框。示例代码如下所示。

```
AlertDialog.Builder builder = new AlertDialog.Builder(this)
        .setTitle("设置字体大小")                //设置标题
        .setIcon(R.mipmap.ic_launcher)
        .setSingleChoiceItems(new String[]{"小号", "默认", "中号", "大号",
                "超大"}, textSize, new DialogInterface.OnClickListener() {
            public void onClick(DialogInterface dialog, int which) {
                textSize = which;
            }
        })
        .setPositiveButton("确定", new DialogInterface.OnClickListener() {
            @Override
            public void onClick(DialogInterface dialog, int which) {
                //为TextView设置在单选对话框中选择的字体大小
                textView.setTextSize(textSizeArr[textSize]);
                dialog.dismiss();                //关闭对话框
            }
```

```
            })//添加"确定"按钮
            .setNegativeButton("取消", new DialogInterface.OnClickListener() {
                @Override
                public void onClick(DialogInterface dialog, int which) {
                    dialog.dismiss();
                }
            });
dialog = builder.create();
dialog.show();
```

图 3.10 是单选对话框示例的运行结果。

图 3.10　单选对话框示例的运行结果

多选对话框的内容区域显示为多选列表。

以下是多选对话框的示例代码,图 3.11 是多选对话框示例的运行结果。

```
AlertDialog.Builder builder = new AlertDialog.Builder(this)
        .setTitle("请添加兴趣爱好!")
        .setIcon(R.mipmap.ic_launcher)
        .setMultiChoiceItems(items, checkedItems,
                new DialogInterface.OnMultiChoiceClickListener() {
                    @Override
                    public void onClick(DialogInterface dialog, int which, boolean
                            isChecked){
                        checkedItems[which] = isChecked;
```

```java
                }
            })
            .setPositiveButton("确定", new DialogInterface.OnClickListener() {
                @Override
                public void onClick(DialogInterface dialog, int which) {
                    StringBuffer stringBuffer = new StringBuffer();
                    for (int i = 0; i <= checkedItems.length - 1; i++) {
                        if (checkedItems[i]) {
                            stringBuffer.append(items[i]).append(" ");
                        }
                    }
                    if (stringBuffer != null) {
                        Toast.makeText(MainActivity.this, "" + stringBuffer,
                                Toast.LENGTH_SHORT).show();
                    }
                    dialog.dismiss();
                }
            })
            .setNegativeButton("取消", new DialogInterface.OnClickListener() {
                @Override
                public void onClick(DialogInterface dialog, int which) {
                    dialog.dismiss();
                }
            });
    dialog = builder.create();
    dialog.show();
```

图3.11 多选对话框示例的运行结果

3.3.3 创建 DatePickerDialog 和 TimePickerDialog 对话框

DatePickerDialog 与 TimePickerDialog 的用法相似，只需要如下两步：

（1）通过构造器创建 DatePickerDialog、TimePickerDialog 实例，调用它们的 show()方法即可将日期选择对话框、时间选择对话框显示出来。

（2）为 DatePickerDialog、TimePickerDialog 绑定监听器，这样可以保证用户通过 DatePickerDialog、TimePickerDialog 选择日期、时间触发监听器，从而通过监听器来获取用户所选择的日期、时间。

3.4 ListView 的使用

ListView 组件主要用于以列表的形式显示数据并响应用户的选择点击事件。

创建 ListView 组件一般需要以下三个元素。

（1）ListView 中每一行的 View。View 可以是系统存在的布局 XML 文件，也可以是自定义的 XML 布局文件。

（2）需要展示的数据。数据既可以来自数组资源，又可以在程序中生成。

（3）连接数据与 ListView 的适配器。

3.4.1 ListView 控件的简单使用

ListView 控件常用方法如表 3.6 所示。

表 3.6 ListView 控件常用方法

方 法	说 明
setAdapter()	为视图设置数据适配器
getAdapter()	获得当前视图适配器
setDivider()	设置元素间的分隔符
setDividerHeight()	设置分隔符的高度
setSelection(int position)	设置视图中的选中项
getMaxScrollAmount()	获得视图的最大滚动数量
setOnItemClickListener(AdapterView.OnItemClickListener listener)	设置 ListView 数据项点击监听器
setOnItemSelectedListener(AdapterView.OnItemSelectedListener listener)	设置数据项选定时的监听器

在 XML 文件的 LinearLayout 中添加 ListView 控件的示例代码如下。

```
<?xml version = "1.0" encoding = "utf - 8"?>
<LinearLayout xmlns:android = "http://schemas.android.com/apk/res/android"
    android:layout_width = "match_parent"
    android:layout_height = "match_parent"
    android:orientation = "vertical">
        <ListView
        android:id = "@ + id/lv"
        android:layout_width = "match_parent"
```

```
            android:layout_height = "wrap_content"/>
</LinearLayout>
```

3.4.2 常用数据适配器

适配器主要用于提供数据转换功能,将源数据转换为目标组件需要的数据格式。Android 可以提供多种数据源,而组件能够识别的数据格式却很单一。因此,需要使用适配器实现数据源与组件之间的数据转换。

Android 针对不同的数据源提供了多种适配器,如 ArrayAdapter、SimpleAdapter、CursorAdapter、BaseAdapter 等。其中,ArrayAdapter 最为简单,它用来绑定一个数组,支持泛型操作。SimpleAdapter 有最好的扩充性,可以自定义各种效果,例如,可以组合 ImageView、Button、CheckBox 等多种组件展示一项数据。CursorAdapter 用来绑定游标得到的数据,它可以看作 SimpleAdapter 与数据库的简单结合,可以方便地把数据库的内容以列表的形式展示出来。BaseAdapter 是一种基础数据适配器,BaseAdapter 类是一个抽象类,用户需要继承该类并实现相应的方法,从而对适配器进行更灵活的操作。

在创建适配器后,可以通过 ListView 对象的 setAdapter() 方法添加适配器,如将继承 BaseAdapter 的 MyBaseAdapter 实例添加到 ListView 中,示例代码如下。

```
//初始化 ListView 控件
mListView = (ListView) findViewById(R.id.lv);
//创建一个 Adapter 的实例
MyBaseAdapter mAdapter = new MyBaseAdapter();
//设置 Adapter
mListView.setAdapter(mAdapter);
```

3.4.3 自定义 ListItem

每个 ListView 控件的列表都是由若干 Item 组成的,例如以下示例,每个 Item 上都显示商品的图片、名称及价格,界面效果如图 3.12 所示。

图 3.12 ListItem 界面效果

3.5 RecyclerView 的使用

在 Android 5.0 之后,Google 提供了用于在有限的窗口范围内显示大量数据的控件 RecyclerView。与 ListView 控件相似,RecyclerView 控件同样是以列表的形式展示数据的,并且数据都是通过适配器加载的。但是 RecyclerView 的功能更加强大,接下来从以下几方面来分析。

(1) 展示效果:RecyclerView 控件可以通过 LayoutManager 类实现横向或竖向的列表效果、瀑布流效果和 GridView 效果,而 ListView 控件只能实现竖直的列表效果。

（2）适配器：RecyclerView 控件使用的是 RecyclerView.Adapter 适配器，该适配器将 BaseAdapter 中的 getView()方法拆分为 onCreateViewHolder()方法和 onBindViewHolder()方法，强制使用 ViewHolder 类，使代码编写规范化，避免了初学者写的代码性能不佳。

（3）复用效果：ListView 控件复用 Item 对象的工作由该控件自己实现，而 ListView 控件复用 Item 对象的工作需要开发者通过 convertView 的 setTag()方法和 getTag()方法进行操作。

（4）动画效果：RecyclerView 控件可以通过 setItemAnimator()方法为 Item 添加动画效果，而 ListView 控件不可以通过该方法为 Item 添加动画效果。

以下逻辑代码是实现 RecyclerView 控件的一个示例，运行结果如图 3.13 所示。

```java
public class MainActivity extends AppCompatActivity {
    private RecyclerView mRecyclerView;
    private HomeAdapter mAdapter;
    private String[] names = { "小猫","哈士奇","小黄鸭","小鹿","老虎"};
    private int[] icons = { R.drawable.cat,R.drawable.siberiankusky,
                R.drawable.yellowduck,R.drawable.fawn, R.drawable.tiger};
    private String[] introduces = {
                "猫,属于猫科动物,分家猫、野猫,是全世界家庭中较为广泛的宠物。",
                "西伯利亚雪橇犬,常见别名哈士奇,昵称为二哈。",
                "鸭的体型相对较小,颈短,一些属的嘴要大些。腿位于身体后方,因而步态蹒跚。",
                "鹿科是哺乳纲偶蹄目下的一科动物。体型大小不等,为有角的反刍类。",
                "虎,大型猫科动物;毛色浅黄或棕黄色,满身黑色横纹;头圆、耳短,耳背面黑色,中央有一白斑甚显著;四肢健壮有力;尾粗长,具黑色环纹,尾端黑色。"
    };
    @Override
    protected void onCreate(Bundle savedInstanceState) {
        super.onCreate(savedInstanceState);
        setContentView(R.layout.activity_main);
        mRecyclerView = (RecyclerView) findViewById(R.id.id_recyclerview);
        mRecyclerView.setLayoutManager(new LinearLayoutManager(this));
        mAdapter = new HomeAdapter();
        mRecyclerView.setAdapter(mAdapter);
    }
    class HomeAdapter extends RecyclerView.Adapter<HomeAdapter.MyViewHolder> {
        @Override
        public MyViewHolder onCreateViewHolder(ViewGroup parent, int viewType) {
            MyViewHolder holder = new MyViewHolder(LayoutInflater.from(MainActivity.
                        this).inflate(R.layout.recycler_item, parent, false));
            return holder;
        }
        @Override
        public void onBindViewHolder(MyViewHolder holder, int position) {
            holder.name.setText(names[position]);
            holder.iv.setImageResource(icons[position]);
            holder.introduce.setText(introduces[position]);
        }
        @Override
        public int getItemCount() {
            return names.length;
        }
```

```
class MyViewHolder extends RecyclerView.ViewHolder {
    TextView name;
    ImageView iv;
    TextView introduce;
    public MyViewHolder(View view) {
        super(view);
        name = (TextView) view.findViewById(R.id.name);
        iv = (ImageView) view.findViewById(R.id.iv);
        introduce = (TextView) view.findViewById(R.id.introduce);
    }
}
```

图 3.13　RecyclerView 控件运行结果

3.6　手势

　　Android 开发中,几乎所有的事件都会和用户进行交互,而最多的交互形式就是手势。目前有多款手机已经支持手写输入,其原理是根据用户输入的内容在预先定义的词库中查找最佳匹配项供用户选择。

　　手势是指用户手指或触摸笔在触摸屏幕上连续碰撞的行为。当用户触摸屏幕时,会产生许多手势,如按下、滑动、弹起等。Android SDK 提供了一个 GestureDetector 类,通过该类可以识别很多复杂的手势。

Android 系统对两种手势提供了支持：一种是在屏幕上从上到下绘制一条线条的简单手势，Android 提供了检测此种手势的监听器；另一种是在屏幕上绘制一个不规则的图形的复杂手势，Android 允许开发商添加此种手势，并提供了相应的 API 识别用户手势。

3.6.1 手势检测

Android 系统提供的 GestureDetector 类用于检测用户的触摸手势，该类内部定义了三个监听接口和一个类，分别是 OnGestureListener 接口、OnDoubleTapListener 接口、OnContextClickListener 接口以及 SimpleOnGestureListener 类。以下代码展示了通过在 raw 文件夹中添加手势文件进行匹配。

```java
public void onCreate(Bundle savedInstanceState) {
    super.onCreate(savedInstanceState);
    setContentView(R.layout.activity_main);
    //在 raw 文件夹中加载手势文件
    library = GestureLibraries.fromRawResource(this, R.raw.gestures);
    GestureOverlayView gesture = (GestureOverlayView) findViewById(R.id.
            gestures);
    gesture.addOnGesturePerformedListener(this); //增加事件监听器
}

@Override
public void onGesturePerformed(GestureOverlayView overlay, Gesture gesture) {
    loadStatus = library.load(); //加载手势库
    if (loadStatus) { //如果手势库加载成功
        //识别绘制的手势，Prediction 是一个相似度对象，集合中的相似度是从高到低进行排列
        ArrayList<Prediction> pres = library.recognize(gesture);
        if (!pres.isEmpty()) {
            Prediction pre = pres.get(0);              //获取相似度最高的对象
            //用整型的数表示百分比,如>40%
            if (pre.score > 4) { //如果手势的相似度分数大于40%,则匹配成功
                Toast.makeText(this, pre.name, Toast.LENGTH_LONG).show();
            } else {
                Toast.makeText(MainActivity.this, "手势匹配不成功",
                        Toast.LENGTH_LONG).show();
            }
        } else {
            Toast.makeText(MainActivity.this, "手势库加载失败",
                    Toast.LENGTH_LONG).show();
        }
    }
}
```

3.6.2 增加手势

Android 系统除了提供手势检测之外，还允许应用程序将用户手势添加到指定文件中，便于后续用户再次绘制该手势时，系统可识别该手势。Android 系统使用 GestureLibrary 来代替手势库，并提供了 GestureLibraries 工具类来创建手势库。

3.7 应用实例：图片浏览器

通过本章学习的内容可以实现一个图片浏览器的 Android 应用程序。此程序主要实现两个功能：其一是通过单选按钮切换界面背景图片；其二是通过"上一张""下一张"两个按钮从图片队列里选择顺序或者逆序展示图片，或者用手指左滑、右滑页面来切换图片。由于需要用到背景图片和需要展示的若干图片，在创建项目之后，需要将所需要的图片导入 res 文件夹的 drawable 文件夹，如图 3.14 所示。

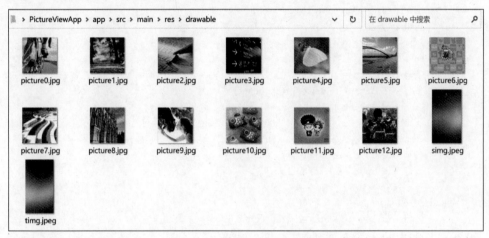

图 3.14　drawable 文件夹

该程序的 XML 界面代码如下所示。

```xml
<?xml version = "1.0" encoding = "utf-8"?>
<RelativeLayout xmlns:android = "http://schemas.android.com/apk/res/android"
    xmlns:app = "http://schemas.android.com/apk/res-auto"
    xmlns:tools = "http://schemas.android.com/tools"
    android:id = "@+id/activity_main"
    android:layout_width = "match_parent"
    android:layout_height = "match_parent"
    android:paddingBottom = "@dimen/activity_vertical_margin"
    android:paddingLeft = "@dimen/activity_horizontal_margin"
    android:paddingRight = "@dimen/activity_horizontal_margin"
    android:paddingTop = "@dimen/activity_vertical_margin"
    tools:context = "com.example.bob50.pictureviewapp.MainActivity"
    android:background = "@drawable/timg">

    <ImageView
        android:layout_width = "wrap_content"
        android:layout_height = "wrap_content"
        app:srcCompat = "@drawable/picture0"
        android:layout_marginTop = "57dp"
        android:id = "@+id/imageView"
```

```
            android:layout_alignParentTop = "true"
            android:layout_centerHorizontal = "true" />

    < Button
          android:text = "上一张"
          android:layout_width = "wrap_content"
          android:layout_height = "wrap_content"
          android:layout_below = "@ + id/imageView"
          android:layout_alignParentStart = "true"
          android:layout_marginStart = "49dp"
          android:layout_marginTop = "94dp"
          android:id = "@ + id/button1" />

    < Button
          android:text = "下一张"
          android:layout_width = "wrap_content"
          android:layout_height = "wrap_content"
          android:layout_alignTop = "@ + id/button1"
          android:layout_alignParentEnd = "true"
          android:layout_marginEnd = "46dp"
          android:id = "@ + id/button2" />

    < RadioGroup
          android:layout_width = "wrap_content"
          android:layout_height = "wrap_content"
          android:layout_alignParentBottom = "true"
          android:layout_centerHorizontal = "true"
          android:layout_marginBottom = "56dp"
          android:id = "@ + id/radioGroup1" >

        < RadioButton
              android:text = "背景 1"
              android:layout_width = "wrap_content"
              android:layout_height = "wrap_content"
              android:id = "@ + id/radioButton1"
              android:layout_weight = "1" />

        < RadioButton
              android:text = "背景 2"
              android:layout_width = "wrap_content"
              android:layout_height = "wrap_content"
              android:id = "@ + id/radioButton2"
              android:layout_weight = "1" />

    </RadioGroup >
</RelativeLayout >
```

该程序的 XML 界面布局如图 3.15 所示。

该程序的逻辑代码如下所示,运行结果如图 3.16 和图 3.17 所示。

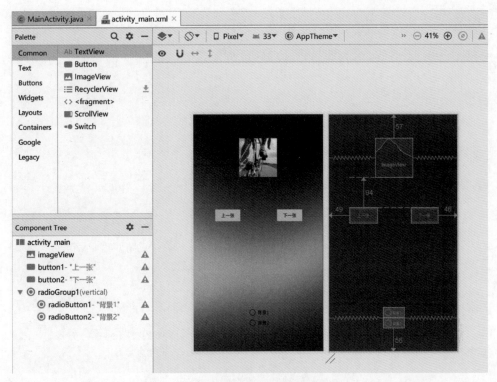

图 3.15　程序的界面布局

```
public class MainActivity extends AppCompatActivity implements View.OnClickListener
{

    private ImageView imageView;
    private Button button1;
    private Button button2;
    private RadioGroup radioGroup1;
    private RelativeLayout relativeLayout;
    //private ArrayList imageList;
    private int[] imageArray;
    private int index = 0;

    private GestureDetector mGestureDetector;
    @Override
    protected void onCreate(Bundle savedInstanceState) {
        super.onCreate(savedInstanceState);
        setContentView(R.layout.activity_main);
        relativeLayout = (RelativeLayout)findViewById(R.id.activity_main);
        imageView = (ImageView) this.findViewById(R.id.imageView);
        button1 = (Button) this.findViewById(R.id.button1);
        button2 = (Button) this.findViewById(R.id.button2);
        button1.setOnClickListener(this);
        button2.setOnClickListener(this);
        radioGroup1 = (RadioGroup)findViewById(R.id.radioGroup1);
        radioGroup1.setOnCheckedChangeListener(new RadioGroup.OnCheckedChangeListener() {
            @Override
            public void onCheckedChanged(RadioGroup radioGroup, int i) {
```

```java
            if(i == R.id.radioButton1)
            {
                relativeLayout.setBackgroundResource(R.drawable.timg);
            }
            if(i == R.id.radioButton2)
            {
                relativeLayout.setBackgroundResource(R.drawable.simg);
            }
        }
    });
    //imageList = new ArrayList<Integer>();
    imageArray = new int[]{R.drawable.picture0, R.drawable.picture0, R.drawable.picture1, R.drawable.picture2,
    R.drawable.picture3, R.drawable.picture4, R.drawable.picture5, R.drawable.picture6, R.drawable.picture7,
    R.drawable.picture8, R.drawable.picture9, R.drawable.picture10, R.drawable.picture11, R.drawable.picture12};

            mGestureDetector = new GestureDetector(this, new GestureDetector.SimpleOnGestureListener(){
                @Override
                public boolean onFling(MotionEvent e1, MotionEvent e2, float velocityX, float velocityY) {
                    //向右滑动表示进入下一页
                    if((e1.getRawX() - e2.getRawX()) > 200)
                    {
                        index++;
                        index = index % 13;
                        Log.d("index:"," " + index);
                        imageView.setImageResource(imageArray[index]);
                        return true;
                    }
                    //向左滑动表示进入上一页
                    if((e2.getRawX() - e1.getRawX()) > 200)
                    {
                        index--;
                        index = (index + 13) % 13;
                        Log.d("index:"," " + index);
                        imageView.setImageResource(imageArray[index]);
                        return true;
                    }
                    return super.onFling(e1, e2, velocityX, velocityY);
                }
            });
    }

    @Override
    public void onClick(View v) {
        switch (v.getId())
        {
            case R.id.button1:
                index--;
                index = (index + 13) % 13;
```

```
                    Log.d("index:"," " + index);
                    this.imageView.setImageResource(this.imageArray[index]);
                    break;
                case R.id.button2:
                    index++;
                    index = index % 13;
                    Log.d("index:"," " + index);
                    this.imageView.setImageResource(this.imageArray[index]);
                    break;
            }
        }

        @Override
        public boolean onTouchEvent(MotionEvent event) {
            return mGestureDetector.onTouchEvent(event);
        }
    }
```

图 3.16　运行结果 1

图 3.17　运行结果 2

3.8 小结

本章主要讲解了 Android 中控件的相关知识,包括简单控件、AlertDialog 对话框、ListView 控件和 RecyclerView 控件以及 ProgressBar 控件等。通过本章的学习,希望初学者能够掌握 Android 控件的基本使用,因为无论创建任何 Android 程序,都可能用到这些控件。

习题 3

1. 简述实现 Button 按钮的点击事件都有哪些方法。
2. 简述 AlertDialog 对话框的创建过程。
3. 简述 ListView 和 RecylerView 的区别。
4. 设计一款移动端计算器的界面。

第 4 章 页面活动单元Activity

本章导图

主要内容

- Activity 的生命周期。
- Activity 的创建。
- Activity 的启动模式。
- Activity 之间数据传递。
- Fragment 应用。

难点

- Activity 的生命周期。
- Activity 之间数据传递。

  Activity 是一个可视化的用户界面,负责创建一个屏幕窗口放置 UI 控件供用户交互。当打开一个应用时,看到的界面就是一个 Activity,当点击一个按钮,跳转到另一个界面时,则又是一个新的 Activity。在应用程序中,Activity 就是一个界面管理员,用户在界面上的操作都是通过 Activity 来实现的。

  Activity 是 Android 应用中最重要、最常见的应用组件,本章将针对 Activity 的相关知识进行详细讲解。

4.1 创建、注册和使用 Activity

  Activity 是 Android 的一个重要组件,它为用户提供一个可视的界面,方便用户进行操作,如拨打电话、照相、发邮件或者是浏览地图等。每个 Activity 都会提供一个可视的窗口界面,在这些界面中还可以添加多个控件,每个控件负责实现不同的功能。

4.1.1 创建 Activity

在 Android 中创建一个 Activity 很简单。在程序的包名处(应用程序已经存在)右击,选择 New→Activity→Empty View Activity 选项,如图 4.1 所示。

图 4.1 创建 Activity

选择 Empty View Activity 选项时,会弹出 New Android Activity 对话框,如图 4.2 所示。

在图 4.2 中,显示了 4 个输入框,分别为 Activity Name、Layout Name、Package name 和 Source Language,这 4 个输入框分别用于输入 Activity 名称、布局名称、包名和所使用的语言。填写完这些信息后,单击 Finish 按钮完成 Activity 的创建。

接下来创建一个 ActivityApp 应用程序,指定包名为 com.example.activityapp,Activity 名为 ActivityExample,根据图 4.1 和图 4.2 中的步骤创建完成的 ActivityExample 的具体代码详见【文件 4-1】。

【文件 4-1】 ActivityExample.java

```
package com.example.activityapp;
import androidx.appcompat.app.AppCompatActivity;
import android.os.Bundle;
public class ActivityExample extends AppCompatActivity {
    @Override
```

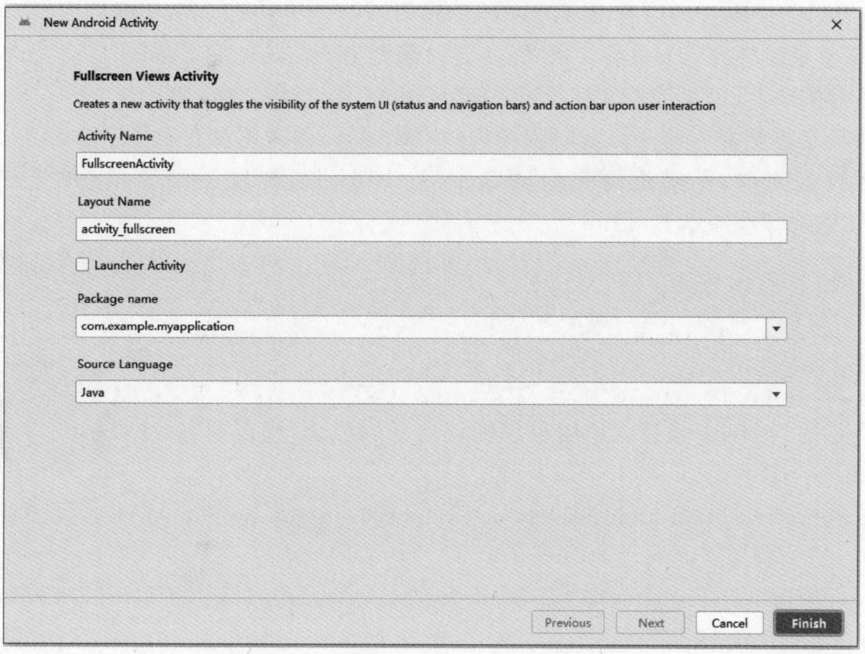

图 4.2　New Android Activity 对话框

```
    protected void onCreate(Bundle savedInstanceState) {
        super.onCreate(savedInstanceState);
        setContentView(R.layout.activity_example);
    }
}
```

4.1.2　注册 Activity

Activity 是 Android 4 大组件之一，Android 规定 4 大组件必须在 AndroidManifest.xml 中注册，如果不注册，则会引起意外错误等问题。

对 Activity 进行注册，具体代码详见【文件 4-2】。

【文件 4-2】　AndroidManifest.xml

```xml
<?xml version = "1.0" encoding = "utf-8"?>
<manifest xmlns:android = "http://schemas.android.com/apk/res/android"
    package = "com.example.activityapp">
    <application
        android:allowBackup = "true"
        android:icon = "@mipmap/ic_launcher"
        android:label = "@string/app_name"
        android:roundIcon = "@mipmap/ic_launcher_round"
        android:supportsRtl = "true"
        android:theme = "@style/Theme.ActivityTest">
        <activity
            android:name = ".ActivityExample"
            android:exported = "true" />
    </application>
</manifest>
```

从文件 4-2 的内容可以看出，Activity 的注册声明位于 application 标签内，这里是通过 activity 标签来对 Activity 进行注册的，不过注册过程是 Android Studio 自动完成的，因此不需要手动维护，可以在清单文件 AndroidManifest.xml 中查看了解。

上面只是注册了 Activity，此时程序仍然不能运行，因为还没有为程序配置主 Activity，即程序运行起来时，不知道要先启动哪个 Activity。而配置主 Activity 的方法就是在 activity 标签内部添加以下内容：

```xml
<intent-filter>
    <action android:name="android.intent.action.MAIN" />
    <category android:name="android.intent.category.LAUNCHER" />
</intent-filter>
```

如果要把 Activity 设置为应用程序默认启动的界面，则需要在<activity>节点中配置<intent-filter>节点。

该节点中的<action android:name="android.intent.action.MAIN"/>表示将当前 Activity 设置为程序最先启动的 Activity。

<category android:name="android.intent.category.LAUNCHER"/>表示让当前 Activity 在桌面上创建图标。

此外，还可以使用 android:label 指定 Activity 中标题栏的内容，标题栏是显示在 Activity 最顶部的，同时给主 Activity 指定的 label 不仅会生成标题栏中的内容，还会成为启动器（Launcher）中应用程序显示的名称，完整的清单具体代码详见【文件 4-3】。

【文件 4-3】 AndroidManifest.xml 完整清单

```xml
<?xml version="1.0" encoding="utf-8"?>
<manifest xmlns:android="http://schemas.android.com/apk/res/android"
    package="com.example.activityapp">
    <application
        android:allowBackup="true"
        android:icon="@mipmap/ic_launcher"
        android:label="@string/app_name"
        android:roundIcon="@mipmap/ic_launcher_round"
        android:supportsRtl="true"
        android:theme="@style/Theme.ActivityTest">
        <activity
            android:name=".ActivityExample"
            android:label="This is ActivityExample"
            android:exported="true">
            <intent-filter>
                <action android:name="android.intent.action.MAIN" />
                <category android:name="android.intent.category.LAUNCHER" />
            </intent-filter>
        </activity>
    </application>
</manifest>
```

4.1.3 使用 Activity

Activity 是用户操作的可视化界面，它为用户提供了一个完成操作指令的窗口，当我们创建完毕 Activity 之后,需要重写 onCreate()方法并调用 setContentView()来完成界面的显示并以此为用户提供交互的入口。

```
package com.example.activityapp
import androidx.appcompat.app.AppCompatActivity
import android.os.Bundle
  class ActivityExample extents AppCompatActivity() {
      override fun onCreate(savedInstanceState: Bundle?) {
          super.onCreate(savedInstanceState)
          setContentView(R.layout.activity_main);
      }
  }
```

可以看到,这里调用了 setContentView()方法来给当前的 Activity 加载一个布局,而在该方法中一般只需要传入一个资源 ID 即可。之前提到,项目中添加的任何资源都会在 R 文件中生成一个相应的资源 ID,这里传入的就是布局文件 activity_main.xml 的 ID。

注:项目中的任何 Activity 都应该重写 onCreate()方法,不过上面的 onCreate()方法是自动生成的。可以看到,上面自动生成的 onCreate()方法很简单,就是调用父类的 onCreate()方法。

4.1.4　Activity 的启动与关闭

1. Activity 的启动:startActivity()

通过调用 startActivity(intent)方法启动 Activity。intent 用来准确地描述要启动的 Activity,当在同一个应用中需要启动另一个 Activity 时,可以使用 intent 指明要启动的 Activity 类。

```
//定义一个 intent,指明要启动的 Activity:SecondActivity
Intent intent = new Intent(MainActivity.this,SecondActivity.class);
//使用 startActivity()方法启动 Activity
startActivity(intent);
```

2. 关闭 Activity:finish()

Activity 可以调用 finish()方法关闭自己,也可以通过调用 finishActivity()方法关闭一个之前启动的独立 Activity。

4.2　Activity 的生命周期和启动模式

在程序开发中,大部分组件都有自己的生命周期,Activity 也有自己的生命周期。所谓生命周期就是一个从创建到销毁的过程。在 Activity 的生命周期中包括 5 种状态,涉及 7 个方法,本节将针对 Activity 的生命周期状态以及生命周期方法进行详细讲解。

4.2.1　Activity 的生命周期状态

Activity 生命周期的 5 种状态分别是启动状态、运行状态、暂停状态、停止状态和销毁状态,其中启动状态和销毁状态是过渡状态,Activity 不会在这两个状态停留。

(1) 启动状态。Activity 的启动状态很短暂,Activity 启动后便会进入运行状态。

(2) 运行状态。Activity 在此状态时处于屏幕最前端,它是可见的、有焦点的,可以与用户进行交互。如点击、长按等事件。即使出现内存不足的情况,Android 也会先销毁栈底的 Activity,来确保当前的 Activity 正常运行。

(3) 暂停状态。在某些情况下,Activity 对用户来说部分可见,但它无法获取焦点,用户对它操作没有响应,此时它处于暂停状态。

(4) 停止状态。当 Activity 完全不可见时,它处于停止状态,但仍然保留着当前的状态和成员信息,如果系统内存不足,那么这种状态下的 Activity 很容易被销毁。

(5) 销毁状态。当 Activity 处于销毁状态时,将被清理出内存。

为了让初学者更好地理解 Activity 的 5 种状态以及不同状态时使用的方法,Google 专门提供了 Activity 的生命周期模型,如图 4.3 所示。

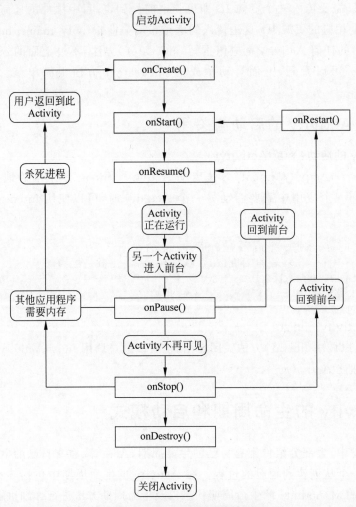

图 4.3 Activity 的生命周期模型

从图 4.3 可以看出,一个 Activity 从启动到关闭,会依次执行 onCreate()→onStart()→onResume()→onPause()→onStop()→onDestroy()方法。

4.2.2 Activity 的生命周期方法

Activity 的生命周期中主要涉及 7 个方法,下面分别对这 7 个方法进行介绍。

(1) onCreate()方法:表示 Activity 正在被创建,这是生命周期的第一个方法。在这个方法中,可以做一些初始化工作,如调用 setContentView()去加载界面布局资源、初始化 Activity 所需数据等。

(2) onStart()方法:表示 Activity 正在被启动,即将开始,但是还没有出现在前台,还无法和用户进行交互。

(3) onRestart()方法:表示 Activity 正在重新启动。一般情况下,当前 Activity 从不可见状态重新变为可见状态时,onRestart()方法就会被调用。

(4) onResume()方法:表示 Activity 已经可见了,并且出现在前台并开始活动。要注意它和 onStart()方法的对比,onStart()方法和 onResume()方法都表示 Activity 已经可见,但是调用 onStart()方法时 Activity 还在后台,调用 onResume()时 Activity 才显示到前台可以交互。

(5) onPause()方法:表示 Activity 正在停止,正常情况下,紧接着 onStop()方法就会被调用。

(6) onStop()方法:表示 Activity 停止,可以做一些回收工作,但不能太耗时。

(7) onDestroy()方法:表示 Activity 即将被销毁,这是 Activity 生命周期中的最后一个回调方法。在这里可以做一些回收工作和最终的资源释放。

为了让初学者更直观地认识 Activity 生命周期,接下来通过一个案例来进行展示。首先创建一个名为 ActivityLife 的应用程序,在 MainActivity、java 中重写 Activity 的生命周期方法,并在每个方法中打印出 Log 以便观察,具体代码详见【文件 4-4】。

【文件 4-4】 MainActivity.java

```java
package com.example.myapplication;
import androidx.appcompat.app.AppCompatActivity;
import android.os.Bundle;
import android.util.Log;
public class MainActivity extends AppCompatActivity {
    @Override
    protected void onCreate(Bundle savedInstanceState) {
        super.onCreate(savedInstanceState);
        setContentView(R.layout.activity_main2);
        Log.i("MainActivityLife","调用了 onCreate()方法");
    }
    @Override
    protected void onStart() {
        super.onStart();
        Log.i("MainActivityLife","调用了 onStart()方法");
    }
    @Override
    protected void onResume() {
        super.onResume();
        Log.i("MainActivityLife","调用了 onResume()方法");
    }
```

```java
@Override
protected void onPause() {
    super.onPause();
    Log.i("MainActivityLife","调用了 onPause()方法");
}
@Override
protected void onStop() {
    super.onStop();
    Log.i("MainActivityLife","调用了 onStop()方法");
}
@Override
protected void onDestroy() {
    super.onDestroy();
    Log.i("MainActivityLife","调用了 onDestroy()方法");
}
@Override
protected void onRestart() {
    super.onRestart();
    Log.i("MainActivityLife","调用了 onRestart()方法");
}
}
```

当第一次运行程序时,在 LogCat 中观察输出日志,可以发现程序启动后依次调用了 onCreate()、onStart()和 onResume()方法。此刻,程序处于运行状态,等待与用户的交互,运行结果如图 4.4 所示。

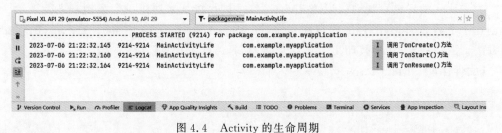

图 4.4 Activity 的生命周期

4.2.3 Activity 的启动模式

通过前面的学习可以发现,Activity 是可以层叠摆放的,每启动一个新的 Activity 就会覆盖在原 Activity 之上,如果点击"返回"按钮,则最上面的 Activity 被销毁,下面的 Activity 重新显示。Activity 之所以能这样显示,是因为 Android 系统是通过任务栈的方式来管理 Activity 实例的。本节将针对 Android 中的任务栈以及 Activity 的启动模式进行详细讲解。

1. Android 中的任务栈

在 Android 系统中,任务栈是一种用来存放 Activity 实例的容器。通常当一个 Android 应用程序启动时,如果当前环境中不存在该应用程序的任务栈,那么系统就会创建一个任务栈。此后,这个应用程序所启动的 Activity 都将在这个任务栈中被管理。需要特别注意的是,一个任务栈中的 Activity 可以来自不同的 App,同一个 App 的 Activity 也可能不在同一个任务栈中。

Activity1 处于栈顶位置,当在 Activity1 中开启 Activity2 时,Activity2 的实例会被压入栈顶的位置。同样,在 Activity2 中开启 Activity3 时,Activity3 的实例也会被压入栈顶的位置。以此类推,无论开启多少个 Activity,最后开启的 Activity 的实例都会被压入栈的顶端,而之前开启的 Activity 虽然"功成身退",却仍然保存在栈中,但活动已经停止。系统会保存 Activity 被停止时的状态。

当用户点击"返回"按钮或者调用 finish()方法时,Activity3 会被弹出栈,Activity2 处于栈顶位置并恢复 Activity2 被保存的界面状态。当然任务栈的这种工作特性并不完美,因此可以给 Activity 设置一些"特权",来打破这种"和谐"的模式。这种"特权"也就是下面将探讨的 Activity 的启动模式。

任务栈的最大特点是先进后出。根据 Activity 当前栈结构中的位置,来决定该 Activity 的状态。正常情况下,Android 任务栈的工作情况如图 4.5 所示。

图 4.5　Android 任务栈的工作情况

2. Activity 的 4 种启动模式

Activity 的启动模式有 4 种,分别是 standard、singleTop、singleTask 和 singleInstance 模式。

1) standard 模式(标准模式)

standard 模式是 Activity 的默认启动方式,每启动一个 Activity 就会在栈顶创建一个新的 Activity 实例。因此,这种启动模式下会存在大量相同的实例。当然,这种模式下也允许存在相同的实例。

2) singleTop 模式(栈顶复用模式)

在某些情况下,会发现使用 standard 模式启动的 Activity 并不合理。例如,当 Activity 已经位于栈顶时,再次启动该 Activity 还需要创建一个新的相同的实例压入任务栈,而不能直接复用之前的 Activity 实例。

在这种情况下,使用 singleTop 模式启动 Activity 更合理,该模式会判断要启动的 Activity 实例是否位于栈顶,如果位于栈顶,则直接复用,否则创建新的实例。实际开发中,浏览器的书签就是采用这种模式。

3) singleTask 模式(栈内复用模式)

使用 singleTop 模式虽然可以很好地解决栈顶重复压入 Activity 实例的问题,但如果要启动的 Activity 没有处于栈顶,则还会在栈中压入多个不相连的 Activity 实例。如果需要 Activity 在栈中有且只有一个实例,则借助 singleTask 模式可以实现。

当 Activity 的启动模式指定为 singleTask 时,每次启动该 Activity 时,系统会首先检查栈中是否存在当前的 Activity 实例,如果存在,则直接使用该 Activity,并将当前 Activity 的实例上面的所有实例全部弹出栈。

4) singleInstance 模式

singleInstance 模式是 4 种启动模式中最特殊的一种，指定 singleInstance 模式的 Activity 会启动一个新的任务栈来管理该 Activity 实例，无论从哪个任务栈中启动该 Activity，该 Activity 实例在整个系统中都只有一个。Android 的桌面使用的就是该模式。

在 singleInstance 模式下，该 Activity 在整个 Android 系统内存中有且只有一个实例，而且该实例单独尊享一个 Task。换句话说，如果 A 应用需要启动的 MainActivity 是 singleInstance 模式，当 A 启动后，系统会为它创建一个新的任务栈，然后 A 单独在这个新的任务栈中，如果此时 B 应用也要激活 MainActivity，由于栈内复用的特性，则不会重新创建，而是两个应用共享一个 Activity 的实例。

4.3 Activity 之间的跳转

在 Android 系统中，每个应用程序通常都由多个界面组成，每个界面就是一个 Activity，在这些界面进行跳转时，实际上也就是 Activity 之间的跳转。Activity 之间的跳转需要用到 Intent（意图）组件，通过 Intent 可以开启新的 Activity 从而实现界面跳转功能。本节将针对 Activity 之间的跳转进行详细讲解。

4.3.1 Intent

Intent 被称为意图，是程序中各组件进行交互的一种重要方式，它不仅可以指定当前组件要执行的动作，还可以在不同组件之间进行数据传递，一般用于启动 Activity、Service 以及发送广播等。根据开启目标组件的方式不同，Intent 被分为显式意图和隐式意图两种类型，接下来分别针对这两种意图进行详细讲解。

1. 显式意图

在通过 Intent 启动 Activity 时需要明确指定激活组件的名称。在程序中，若是需要在本应用中启动其他 Activity，则可以使用该意图来启动 Activity。

显示意图启动 Activity 的代码：

```
Intent intent = new Intent(this,Activity02.class);
startActivity(intent);
```

可以看出显式意图可以直接通过名称开启指定的目标组件，具体可通过其构造方法 Intent(Context packageContext, Class <?> cls)来实现。其中，第 1 个参数为 Context 表示当前的 Activity 对象，这里使用 this 即可；第 2 个参数 Class 表示要启动的目标 Activity。

2. 隐式意图

如果没有明确指定组件的 Intent 则使用隐式意图，此时系统会根据该意图中的动作（Action）、类别（Category）和数据（Uri 和数据类型）寻找合适的组件。

隐式意图相比显式意图来说更为抽象，它并没有明确指定要开启哪个目标组件，而是通过指定 action 和 category 等属性信息，系统根据这些信息进行分析，然后寻找目标 Activity。其示例代码如下。

```
Intent intent = new Intent();                //设置action动作
intent.setAction("com.example.START_ACTIVITY");
startActivity(intent);
```

在上述代码中，只指定了action，并没有指定category，这是因为在目标Activity的清单文件中配置的category是一个默认值，在调用startActivity()方法时，自动将这个category添加到Intent中。

很显然，只有上述代码还不能开启指定的Activity，还需在目标Activity的清单文件中配置<intent-filter>，指定当前Activity能响应的action和category，示例代码如下：

```
<activity android:name=".Activity02">
        <intent-filer>
        <!--设置action属性，需要在代码中根据所设置的name打开指定的组件-->
            <action android:"name="com.example.START_ACTIVITY"/>
            <category ?android:name="android.intent.category.DEFAULT"/>
        <intent-filter>
</activity>
```

注：显示意图开启组件时必须指定组件的名称，而且显示意图一般只在本应用程序切换组件时使用。

3. 实战演练——打开浏览器

通过上面的学习知道，显式意图非常简单，而隐式意图还有很多内容需要学习。使用隐式意图不仅可以启动自己程序中的Activity，还可以启动其他程序中的Activity，这使得程序之间可以共享某些功能。例如，一个程序需要展示网页，而这时又没有必要自己再写一个浏览器，就可以直接调用系统中的浏览器打开网页。通过隐式意图来实现打开系统浏览器的具体步骤如下。

（1）创建应用程序。

创建一个名为OpenBrowserApp的应用程序，指定包名为cn.example.openbrowser，设计用户交互界面，预览效果如图4.6所示。

程序主界面对应的布局具体代码详见【文件4-5】。

【文件4-5】 activity_main.xml

```
<?xml version="1.0" encoding="utf-8"?>
<RelativeLayout xmlns:android="http://schemas.android.com/apk/res/android"
    xmlns:app="http://schemas.android.com/apk/res-auto"
    xmlns:tools="http://schemas.android.com/tools"
    android:layout_width="match_parent"
    android:layout_height="match_parent"
    tools:context=".MainActivity">
    <Button
        android:layout_width="wrap_content"
        android:layout_height="wrap_content"
        android:textSize="25dp"
        android:text="打开浏览器百度页面"
        android:id="@+id/btnopen"
        android:layout_marginTop="100dp"
```

```
                android:layout_centerHorizontal = "true"/>
</RelativeLayout>
```

(2) 编写界面交互程序。

在 MainActivity 中,通过隐式意图打开系统的浏览器并访问百度页面,具体代码详见【文件 4-6】。

【文件 4-6】 MainActivity.java

```java
import androidx.appcompat.app.AppCompatActivity;
import android.content.Intent;
import android.net.Uri;
import android.os.Bundle;
import android.view.View;
import android.widget.Button;
public class MainActivity extends AppCompatActivity {
    private Button btnOpenBrower;
    @Override
    protected void onCreate(Bundle savedInstanceState) {
        super.onCreate(savedInstanceState);
        setContentView(R.layout.activity_main);
        btnOpenBrower = findViewById(R.id.btnOpenBrower);
        btnOpenBrower.setOnClickListener(new View.OnClickListener() {
            @Override
            public void onClick(View view) {
                Intent intent = new Intent();
                intent.setAction("android.intent.action.VIEW");
                intent.setData(Uri.parse("http://www.baidu.com"));
                startActivity(intent);
            }
        });
    }
}
```

点击"打开浏览器百度页面"按钮,运行结果如图 4.7 所示。

图 4.6 程序主界面

图 4.7 运行结果

4.3.2 Activity 的数据传递

1. 数据传递

在 Activity 启动时传递数据非常简单,因为 Intent 提供了一系列重载的 putExtra (String name, String value)方法,通过该方法可以将要传递的数据暂存到 Intent 中,当启动另一个 Activity 之后,只需将这些数据从 Intent 中取出即可。

例如将 Activity01 中的字符串传递到 Activity02 中,示例代码如下。

观看视频

```
Intent intent = new Intent(this, Activity02.class);
intent.putExtra("data", "Hello Activity02");
startActivity(intent);
```

在上述代码中,通过 Intent 开启 Activity02,用 putExtra()方法传递了一个字符串"Hello Activity02"。putExtra()方法中接收两个参数:第 1 个参数是键,用于后面从 Intent 中取值;第 2 个参数是传递的数据内容。

接下来在 Activity02 中取出传递过来的数据,示例代码如下。

```
Intent intent = getIntent();
String data = intent.getStringExtra("data");
```

在上述代码中,通过 getIntent()方法获取 Intent 对象,然后调用 getStringExtra(String name)方法,根据传入的键值,取出相应的数据。由于这里传递的是字符串类型数据,因此使用 getStringExtra()方法获取传递的数据。如果传递的数据是整数类型,则使用 getIntExtra()方法,以此类推。

2. 实战演练——注册用户信息

上面讲解了如何使用 Intent 在 Activity 中进行数据传递,由于在实际开发中数据传递使用频率较高,因此接下来通过一个注册用户信息的案例展示 Activity 中的数据传递,具体步骤如下。

(1) 创建程序。创建一个名为 UserRegistApp 的应用程序,指定包名为 com.examplet.userregistapp 并导入图片资源。

(2) 设计用户信息注册界面,预览效果如图 4.8 所示,对应的布局文件代码详见【文件 4-7】。

(3) 设计用户信息显示界面,预览效果如图 4.9 所示,对应的布局文件代码详见【文件 4-8】。

(4) 编写主程序及"注册"按钮事件代码,代码详见【文件 4-9】。

(5) 编写显示用户信息程序,代码详见【文件 4-10】。

【文件 4-7】 activity_main.xml 用户注册布局文件

```
<?xml version = "1.0" encoding = "utf-8"?>
<RelativeLayout xmlns:android = "http://schemas.android.com/apk/res/android"
    xmlns:app = "http://schemas.android.com/apk/res-auto"
    xmlns:tools = "http://schemas.android.com/tools"
```

图 4.8 注册界面　　　　图 4.9 显示信息界面

```
        android:layout_width = "match_parent"
        android:layout_height = "match_parent"
        android:background = "@drawable/bg"
        tools:context = ".MainActivity">
    < ImageView
            android:layout_width = "80dp"
            android:layout_height = "80dp"
            android:src = "@drawable/head"
            android:layout_centerHorizontal = "true"
            android:layout_marginTop = "100dp"
            android:id = "@ + id/imgpic"/>
    < EditText
            android:layout_width = "300dp"
            android:layout_height = "wrap_content"
    android:hint = "请输入用户名"
        android:textSize = "25dp"
        android:textColor = "#ffffff"
        android:layout_below = "@id/imgpic"
        android:layout_centerHorizontal = "true"
        android:layout_marginTop = "20dp"
        android:id = "@ + id/edtusername"/>
        < EditText
            android:layout_width = "300dp"
            android:layout_height = "wrap_content"
            android:hint = "请输入密码"
            android:textSize = "25dp"
            android:layout_below = "@id/edtusername"
            android:layout_centerHorizontal = "true"
            android:layout_marginTop = "20dp"
            android:inputType = "numberPassword"
            android:id = "@ + id/edtpassword"/>
```

```xml
        < Button
            android:layout_width = "200dp"
            android:layout_height = "60dp"
            android:background = "#70AAA5"
            android:text = "注    册"
            android:textSize = "25dp"
            android:layout_centerHorizontal = "true"
            android:layout_below = "@id/edtpassword"
            android:layout_marginTop = "20dp"
            android:id = "@ + id/btnregister"/>
</RelativeLayout >
```

上述布局采用 RelativeLayout，有一个 ImageView 控件显示头像，两个 EditText 控件用于输入用户名和密码，一个 Button 控件用于执行注册命令。

【文件 4-8】 activity_show.xml 用户信息显示界面布局文件

```xml
<?xml version = "1.0" encoding = "utf - 8"?>
< RelativeLayout xmlns:android = "http://schemas.android.com/apk/res/android"
    xmlns:app = "http://schemas.android.com/apk/res - auto"
    xmlns:tools = "http://schemas.android.com/tools"
    android:layout_width = "match_parent"
    android:layout_height = "match_parent"
    android:background = "@drawable/bj2"
    tools:context = ".ShowActivity">
    < ImageView
        android:layout_width = "80dp"
        android:layout_height = "80dp"
        android:layout_centerHorizontal = "true"
        android:layout_marginTop = "100dp"
        android:src = "@drawable/tx" />
    < TextView
        android:id = "@ + id/txtusername"
        android:layout_width = "wrap_content"
        android:layout_height = "wrap_content"
        android:layout_marginLeft = "60dp"
        android:layout_marginTop = "220dp"
        android:text = "用户名  :"
        android:textColor = "#ffffff"
        android:textSize = "25dp" />
    < TextView
        android:id = "@ + id/txtpassword"
        android:layout_width = "wrap_content"
        android:layout_height = "wrap_content"
        android:layout_below = "@id/txtusername"
        android:layout_marginLeft = "60dp"
        android:layout_marginTop = "20dp"
        android:text = "密码   :"
        android:textColor = "#ffffff"
        android:textSize = "25dp" />
</RelativeLayout >
```

上述布局采用 RelativeLayout，有两个 TextView 控件，分别用于显示用户名和密码。

【文件 4-9】 MainActivity.java 主程序及"注册"按钮事件

```java
package com.example.userregistapp;
import androidx.appcompat.app.AppCompatActivity;
import android.content.Intent;
import android.os.Bundle;
import android.view.View;
import android.widget.Button;
import android.widget.EditText;
public class MainActivity extends AppCompatActivity {
    private EditText edtUsername;
    private EditText edtPassword;
    private Button btnRegister;
        @Override
        protected void onCreate(Bundle savedInstanceState) {
            super.onCreate(savedInstanceState);
            setContentView(R.layout.activity_main);
            edtUsername = findViewById(R.id.edtusername);
            edtPassword = findViewById(R.id.edtpassword);
            btnRegister = findViewById(R.id.btnregister);
            btnRegister.setOnClickListener(new View.OnClickListener() {
                @Override
                public void onClick(View view) {
                    Intent intent = new Intent(MainActivity.this,ShowActivity.class);
                    intent.putExtra("username",edtUsername.getText().toString());
                    intent.putExtra("password",edtPassword.getText().toString());
                    startActivity(intent);
                }
            });
        }
}
```

上述代码中,点击"注册"时,Intent 作为载体进行数据的传递。接下来在 ShowActivity 中编写代码,用于接收数据并显示。

【文件 4-10】 ShowActivity.java 接收并显示数据

```java
package com.example.userregistapp;
import androidx.appcompat.app.AppCompatActivity;
import android.content.Intent;
import android.os.Bundle;
import android.widget.TextView;
public class ShowActivity extends AppCompatActivity {
    private TextView txtUsername;
    private TextView txtPassword;
        @Override
        protected void onCreate(Bundle savedInstanceState) {
            super.onCreate(savedInstanceState);
            setContentView(R.layout.activity_show);
            txtUsername = findViewById(R.id.txtusername);
            txtPassword = findViewById(R.id.txtpassword);
            Intent intent = getIntent();
            String username = intent.getStringExtra("username");
```

```
        String password = intent.getStringExtra("password");
        txtUsername.setText("用户名是:" + username);
        txtPassword.setText("密码是:" + password);
    }
}
```

上述代码中,通过 getIntent()方法获取到 Intent 对象,然后通过该对象的 getStringExtra()方法得到输入的用户名和密码,并将得到的数据绑定到 TextView 控件中进行显示。

可以看到,在注册界面中输入的数据成功地传递给了数据显示页面进行了显示,这就是使用 Intent 在 Activity 之间进行数据传递的用法。

4.3.3 Activity 数据回传

在 Activity 中,使用 Intent 既可以将数据传给下一个 Activity,还可以将数据回传给上一个 Activity。Activity 中提供了一个 startActivityForResult(Intent intent,int requestCode)方法,该方法也用于启动 Activity,并且这个方法可以在当前 Activity 销毁时返回一个结果给上一个 Activity。这种功能在实际开发中很常见,例如发微信朋友圈时,进入图库选择好照片后,会返回到发表状态页面并带回所选的图片信息。

startActivityForResult(Intent intent,int requestCode)方法接收两个参数:第 1 个参数是 Intent 对象;第 2 个参数是请求码,用于判断数据的来源,输入一个唯一值即可。使用该方法在 Activity01 中开启 Activity02 的示例代码如下。

```
Intent intent = new Intent(thisActivity02.class);
startActivityForResult(intent,1);
```

接下来在 Activity02 中添加返回数据的示例代码,具体如下。

```
Intent intent = new Intent();
intent.putExtra("data","Hello Activity01");
setResult(1,intent);
```

在上述代码中,构建了一个 Intent 对象,然后调用 setResult(int resultCode,Intent data)方法向上一个 Activity 回传数据,其中第 1 个参数用于向 Activity01 返回处理结果,一般使用"0"或"1",第 2 个参数是把带有数据的 Intent 传递回去。

由于使用 starActivityForResult()方法启动 Activity02,在 Activity02 被销毁之后会回调 Activity01 的 onActivityResult()方法,因此需要在 Activity01 中重写该方法来得到返回的数据,示例代码如下。

```
protected void onActivityResult(int requestCode, int resultCode, Intent data){
    super.onActivityResult(requestCode,resultCode,data);
    if(requestCode == 1){
        if(resultCode == 1){
            String string == data.getStringExtra("data");
        }
    }
}
```

在 Activity01 中，通过实现 onActivityResult(int requestCode, int resultCode, intent data)方法来获取返回的数据。该方法有 3 个参数：第 1 个参数 requestCode 表示在当前 Activity01 启动 Activity02 时传递的请求码；第 2 个参数 resultCode 表示在 Activity02 返回数据时传入结果码；第 3 个参数 data 表示携带返回数据的 Intent。

需要注意的是，在一个 Activity 中很可能调用 startActivityForResult()方法启动多个不同的 Activity，每一个 Activity 返回的数据都会回调到 onActivityResult()这个方法中。因此，首先要做的就是通过检查 requestCode 的值来判断数据来源，确定数据是从 Activity02 返回的，然后通过 resultCode 的值判断数据处理结果是否成功，最后从 data 中取出数据，这样就完成了 Activity 数据回传的功能。

4.4 Fragment 控件

Fragment（碎片）是一种可以嵌入 Activity 中的 UI 片段，与 Activity 非常相似，不仅包含布局，同时也具有自己的生命周期。Fragment 是专门针对大屏幕移动设备而推出的，它能让程序更加合理地利用屏幕空间，因此在平板电脑上应用广泛，本节将针对 Fragment 进行详细讲解。

4.4.1 Fragment 简介

Fragment 是 Android 3.0 以后引入的一个新的 API，它出现的初衷是为了适应大屏幕的平板电脑，当然现在它仍然是平板电脑 App UI 设计的宠儿，而且普通手机开发也会加入这个 Fragment，我们可以把它看成一个小型的 Activity，又称 Activity 片段。使用 Fragment 可以把屏幕划分成几块，然后进行分组，进行一个模块化的管理，从而可以更加方便地在运行过程中动态地更新 Activity 的用户界面。

Fragment 是如何利用大屏幕空间的呢？下面看一个示例。在新闻界面中使用 RecyclerView 控件展示一组新闻标题，当单击新闻标题时就会在另一个界面展示新闻内容。如果在手机中设计这个界面，就会将新闻标题列表放在一个 Activity 中，将新闻内容放在另一个 Activity 中。如果在平板电脑上也这样设计，新闻标题就会被拉长填充屏幕，通常情况下新闻标题不会很长，这样就会导致屏幕有大量空白区域，因此，在平板电脑上更好的设计方式是将新闻标题列表界面和新闻内容界面分开放在两个 Fragment 中，接下来通过一个图例进行展示，如图 4.10 所示，Fragment 分别对应手机与平板电脑间不同情况的处理图。

从图 4.10 可以看出，在普通手机上展示新闻列表和新闻内容各需要一个 Activity，由于普通手机尺寸较小，因此一屏展示新闻列表，一屏展示新闻内容是合理的。如果在平板电脑上这样做就太浪费屏幕空间了，因此通常在平板电脑上都是用一个 Activity 包含两个 Fragment，其中一个 Fragment 用于展示新闻列表，另一个用于展示新闻内容。

4.4.2 Fragment 的生命周期

Fragment 必须是依存于 Activity 而存在的，因此 Activity 的生命周期会直接影响 Fragment 的生命周期。Fragment 状态与 Activity 类似，存在如下 4 种状态。

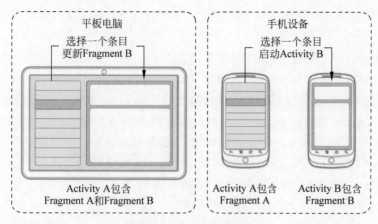

图 4.10 Fragment 示意

运行：当前 Fragment 位于前台，用户可见，可以获得焦点。
暂停：其他 Activity 位于前台，该 Fragment 依然可见，只是不能获得焦点。
停止：该 Fragment 不可见，失去焦点。
销毁：该 Fragment 被完全删除，或该 Fragment 所在的 Activity 被结束。
Fragment 的生命周期与 Activity 的生命周期十分相似，如图 4.11 所示。

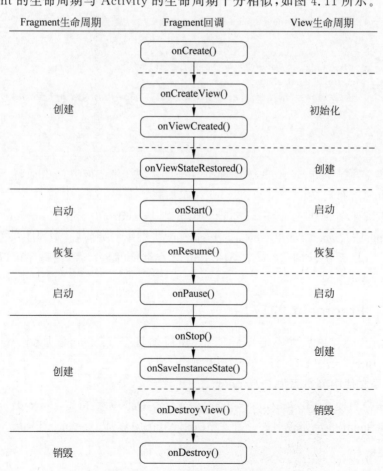

图 4.11 Fragment 的生命周期

可以看出,Activity 中的生命周期方法,Fragment 中基本都有,但是 Fragment 比 Activity 多几个方法。

- onCreate():创建 Fragment 时被回调。
- onCreateView():每次创建、绘制该 Fragment 的 View 组件时回调该方法。
- onStart():启动 Fragment 时被回调,此时 Fragment 可见。
- onResume():恢复 Fragment 时被回调,获取焦点时回调。
- onPause():暂停 Fragment 时被回调,失去焦点时回调。
- onStop():停止 Fragment 时被回调,Fragment 不可见时回调。
- onDestroyView():销毁与 Fragment 有关的视图,但未与 Activity 解除绑定。
- onDestroy():销毁 Fragment 时被回调。

Fragment 具有与 Activity 很相似的生命周期,依存于 Activity 而存在,因此 Activity 的生命周期会直接影响 Fragment 的生命周期。

4.4.3 Fragment 的创建

Fragment 的创建与 Activity 创建类似,创建 Fragment 时必须创建一个类继承自 Fragment。创建一个 Fragment 的示例代码如下。

```java
public class NewsListFragment extends Fragment {
    @Nullable
    @Override
    public View onCreateView(@NonNull LayoutInflater inflater, @Nullable ViewGroup container, @Nullable Bundle savedInstanceState) {
        View view = inflater.inflate(R.layout.fragment,container,false);
        return view;
    }
}
```

上述代码重写了 fragment 的 onCreateView()方法,并在该方法中通过 layoutInflater 的 inflate()方法将布局文件 fragment.xml 动态加载到 Fragment 中。

Android 系统中提供两个 Fragment 类,分别是 android.app.Fragment 和 android.support.v4.app.Fragment。如果 NewsListFragment 类继承的是前者,则程序只能兼容 3.0 版本以上的 Android 系统,如果 NewsListFragment 类继承的是后者,则程序可以兼容 1.6 版本以上的 Android 系统。

观看视频

4.4.4 Fragment 的应用

Fragment 创建完成后并不能单独使用,还需要将 Fragment 添加到 Activity 中。在 Activity 中添加 Fragment 有两种方式,具体如下。

1. 在布局文件中静态添加 Fragment

这种方法需要使用<fragment></fragment>标签,该标签指定 android:name 属性,其属性值为 Fragment 的全路径名称。在 LinearLayout 中添加 NewsListFragment 的示例代码如下。

```xml
<LinearLayout xmlns:android = "http://schemas.android.com.apk/res/android"
    xmlns:tools = "http://schemas.android.com./tools"
    android:layout_width = "match_parent"
    android:layout_height = "match_parent"
    tools:context = ".MainActivity">
<fragment
    android:name = "com.example.NewsListFragment"
    android:id = "@+id/newslist"
    android:layout_width = "match_parent"
    android:layout_height = "match_parent"/>
</Linearlayout>
```

2．在 Activity 中动态加载 Fragment

具体步骤如下。

（1）创建一个 Fragment 的实例对象。

（2）获取 FragmentManager（Fragment 管理器）的实例。

（3）开启 FragmentTransactiion（事务）。

（4）向 Activity 的布局容器（一般为 FrameLayout）中添加 Fragment。

（5）通过 commit（）方法提交事务。

示例代码如下。

```java
public class MAinActivity extends Activity{
    @SuppressLint("NewApi")
    @Override
    protected void onCreate(Bundle savedInstanceSrate){
        super.onCrete(savedInstanceState);
        setContentView(R.layout.activity_main);
        NewsListFragment fragment = new NewsListFragment();
        FragmentManager fm = getFragmentManager();
        //获取 FragmentTransaction 实例
        FragmentTransaction beginTransaction = fm.beginTransact();
        beginTransaction.replace(R.id.11,fragment);         //添加一个 Fragment
        beginTransac.commit();                              //提交事务
    }
```

需要注意的是，调用 replace（）方法将 Fragment 添加到 Activity 布局中时，需要导入 androidx.fragment.app.Fragment 和 androidx.fragment.app.FragmentManager。

4.5 应用实例：餐厅点餐

为了让初学者更好地掌握 Fragment 的使用，接下来我们以图 4.12 所示的餐厅点餐界面为例，演示如何在一个 Activity 中展示两个 Fragment（一个 Fragment 用于展示餐品分类，另一个 Fragment 用于展示餐品信息），并实现 Activity 与 Fragment 通信。具体步骤如下。

1．创建项目工程

创建一个名为 OrderFoodApp 的应用程序，指定包名为 com.example.orderfood。

图 4.12　餐厅点餐界面

2．导入图片资源

将项目中所需要的图片资源导入 drawable 文件夹中。

3．创建主界面布局

在 res/layout 文件夹中的 activity_main.xml 文件中添加两个 FrameLayout 控件，分别用于显示左侧餐品分类列表和右侧餐品信息列表，具体代码详见【文件 4-11】。

4．创建两个 Fragment 的布局文件

由于本实例需要实现一个 Activity 中展示两个 Fragment 的效果，其中一个 Fragment 用于是显示餐品分类列表，另一个用于显示对应的餐品信息，因此需要在 res/layout 文件夹中分别创建布局文件 fragment_menu.xml 和 fragment_food.xml。

展示分类列表的布局文件 fragment_menu.xml 中放置了一个 ListView 控件用于显示餐品分类，具体代码详见【文件 4-12】。

展示餐品信息的布局文件 fragment_food.xml 中放置了一个 ListView 控件用于显示对应的餐品信息列表，具体代码详见【文件 4-13】。

【文件 4-11】 activity_main.xml 布局文件

```xml
<?xml version = "1.0" encoding = "utf-8"?>
<LinearLayout xmlns:android = "http://schemas.android.com/apk/res/android"
    xmlns:tools = "http://schemas.android.com/tools"
    android:layout_width = "match_parent"
    android:layout_height = "match_parent"
    android:orientation = "horizontal"
    tools:context = ".MainActivity">
    <FrameLayout
        android:id = "@+id/menucontent"
        android:layout_weight = "1.5"
        android:layout_width = "0dp"
        android:layout_height = "match_parent">
    </FrameLayout>
    <FrameLayout
        android:id = "@+id/foodcontent"
        android:layout_weight = "3"
        android:layout_width = "0dp"
        android:layout_height = "match_parent">
    </FrameLayout>
</LinearLayout>
```

上面主布局文件里面有两个 FrameLayout 控件，分别用于显示左侧分类和右侧餐品信息。

【文件 4-12】 fragment_menu.xml 布局文件

```xml
<?xml version = "1.0" encoding = "utf-8"?>
<LinearLayout
```

```
    xmlns:android = "http://schemas.android.com/apk/res/android"
    android:layout_width = "match_parent"
    android:orientation = "horizontal"
    android:layout_height = "match_parent">
  <ListView
    android:listSelector = "#CFC3C3"
    android:background = "#959090"
    android:divider = "#CFC3C3"
    android:id = "@+id/menulist"
    android:layout_width = "match_parent"
    android:layout_height = "wrap_content" />
</LinearLayout>
```

该布局里面有一个 ListView 控件,id="@+id/menulist"用于显示分类列表。

【文件 4-13】 fragment_food.xml 布局文件

```
<?xml version = "1.0" encoding = "utf-8"?>
<LinearLayout
    xmlns:android = "http://schemas.android.com/apk/res/android"
    android:layout_width = "match_parent"
    android:orientation = "vertical"
    android:layout_height = "match_parent">
  <ListView
    android:fadingEdge = "vertical"
    android:divider = "@color/white"
    android:id = "@+id/foodlist"
    android:layout_width = "match_parent"
    android:layout_height = "wrap_content"/>
</LinearLayout>
```

该布局里面有一个 ListView 控件,id="@+id/foodlist"用于显示相应的餐品列表。

5. 创建餐品列表的 Item 布局文件

由于展示餐品列表的界面用到了 ListView 控件,因此需要为该列表控件创建一个 Item 布局,界面效果如图 4.13 所示。

在 res/layout 文件夹中创建一个 Item 界面的布局文件 food_item.xml,在该界面中放置 1 个 ImageView 控件用于显示菜品图片,3 个 TextView 控件用于显示餐品信息,具体代码详见【文件 4-14】。

图 4.13 Item 布局效果

【文件 4-14】 food_item.xml 布局文件

```
<?xml version = "1.0" encoding = "utf-8"?>
<LinearLayout xmlns:android = "http://schemas.android.com/apk/res/android"
    android:layout_width = "match_parent"
    android:layout_height = "match_parent"
    android:orientation = "horizontal"
    android:paddingLeft = "8dp"
    android:paddingTop = "8dp">
  <ImageView
      android:id = "@+id/food_icon"
      android:layout_width = "100dp"
```

```xml
            android:layout_height = "100dp"
            android:paddingLeft = "10dp"
            android:paddingTop = "20dp" />
    <LinearLayout
        android:layout_width = "match_parent"
        android:layout_height = "100dp"
        android:orientation = "vertical">
        <TextView
            android:id = "@ + id/food_name"
            android:layout_width = "match_parent"
            android:layout_height = "wrap_content"
            android:paddingLeft = "10dp"
            android:text = "卤蛋螺蛳粉"
            android:textSize = "18dp" />
        <TextView
            android:id = "@ + id/estimate"
            android:layout_width = "match_parent"
            android:layout_height = "wrap_content"
            android:layout_marginTop = "6dp"
            android:paddingLeft = "15dp"
            android:text = "月售:999"
            android:textSize = "12dp" />
        <TextView
            android:id = "@ + id/food_price"
            android:layout_width = "100dp"
            android:layout_height = "wrap_content"
            android:paddingLeft = "10dp"
            android:text = "￥20.05"
            android:textColor = "#CD1E1E"
            android:textSize = "15dp" />
    </LinearLayout>
</LinearLayout>
```

6. 创建 MenuFragment 类

在 com.example.orderfood 包中创建一个 MenuFragment 类继承自 Fragment,在该类中获取界面控件并将餐品分类数据显示到控件上,具体代码详见【文件 4-15】。

【文件 4-15】 MenuFragment 类文件

```java
public class MenuFragment extends Fragment {
    private String[] tips = {"推荐","折扣","进店必买","人气精选","店内招牌","热销单人餐","多人餐","加料区","饮料区","店内环境","有事联系"};
    private View mView;
    private ListView mListView;
    @Nullable
    @Override
    public View onCreateView(LayoutInflater inflater, @Nullable ViewGroup container,
                             Bundle savedInstanceState) {
        mView = inflater.inflate(R.layout.fragment_menu, container, false);
        if (mView != null) {
            initView();
```

```
                }
                mListView.setOnItemClickListener(new AdapterView.OnItemClickListener() {
                    @Override
                    public void onItemClick(AdapterView<?> parent, View view, int position,
                                        long id) {
                        FoodFragmentfoodFragment = (FoodFragment)(getActivity())
                            .getSupportFragmentManager().findFragmentById(R.id.foodcontent);
                        foodFragment.setType(position);
                    }
                });
                return mView;
            }
            private void initView() {
                mListView = mView.findViewById(R.id.menulist);
                ArrayAdapter<String> adapter = new ArrayAdapter<>(
                                    getActivity(), R.layout.item, R.id.text, tips);
                mListView.setAdapter(adapter);
            }
        }
```

上述代码中,通过 inflate()方法加载布局文件 fragment_menu.xml,通过 setOnItemClickListener()方法为分类列表中的 Item 添加点击事件的监听器,在该监听器中重写 onItemClick()方法,通过 getActivity()方法获取 Activity 的实例对象,接着通过该对象的 getSupportFragmentManager()方法获取 FragmentManager 的实例对象,通过 findFragmentById()方法获取到 FoodFragment 对象,最后通过 initView()方法刷新列表数据,将相对应的餐品信息显示到界面控件上。

7. 创建 FoodFragment 类

在 com.example.orderfood 包中创建一个 FoodFragment 类继承自 Fragment。在该类中实现显示餐品列表的信息,具体代码详见【文件 4-16】。

【文件 4-16】 FoodFragment 类文件

```
public class FoodFragment extends Fragment {
    //推荐
    private int food_icons0[] = {R.drawable.zsb, R.drawable.rxb, R.drawable.jzb,
        R.drawable.nwb, R.drawable.jcb, R.drawable.jpb, R.drawable.xjmg, R.drawable.pic};
    private String food_name0[] = {"猪手煲", "肉蟹煲", "鸡爪煲", "牛蛙煲", "鸡翅煲",
                                    "牛排煲", "小鸡炖蘑菇", "虾煲"};
    //折扣
    private int food_icons1[] = {R.drawable.pic, R.drawable.pic};
    private String food_name1[] = {"青菜", "白菜"};
    //进店必买
    private int food_icons2[] = {R.drawable.pic, R.drawable.pic, R.drawable.pic,
                            R.drawable.pic, R.drawable.pic};
    private String food_name2[] = {"螺蛳粉", "云吞", "白粥", "皮蛋瘦肉粥", "玉米粥"};
    //人气精选
    private int food_icons3[] = {R.drawable.pic, R.drawable.pic};
    private String food_name3[] = {"乌鸡汤", "排骨汤"};
    private String food_evaluation[] = {"月售:999", "月售:999 ", "月售:999", "月售:999",
```

```java
        "月售:999", "月售:999", "月售:999 ", "月售:999", "月售:999", "月售:999", "月售:999", "月
售:999", "月售:999"};
    private String food_price[] = {"¥15.99", "¥12.99", "¥22.99", "¥16.99", "¥82.99",
"¥56.99", "¥30.99", "¥12.99", "¥12.99", "¥12.99", "¥12.99", "¥12.99", "¥12.99"};
    private View mView;
    private ListView mListView;
    public MyAdapter mAdapter;
    @Nullable
    @Override
    public View onCreateView(LayoutInflater inflater, @Nullable ViewGroup container,
                Bundle savedInstanceState) {
        mView = inflater.inflate(R.layout.fragment_food, container, false);
        initView();
        return mView;
    }
    private void initView() {
        mListView = mView.findViewById(R.id.foodlist);
        mAdapter = new MyAdapter(0);
        mListView.setAdapter(mAdapter);
    }
    public void setType(int type) {
        mAdapter = null;
        mAdapter = new MyAdapter(type);
        mListView.setAdapter(mAdapter);
    }
    private class MyAdapter extends BaseAdapter {//配置适配器
        private int TYPE;
        public MyAdapter(int type) {
            TYPE = type;
        }
        @Override
        public int getCount() {
            switch (TYPE) {
                case 0:
                    return food_icons0.length;
                case 1:
                    return food_icons1.length;
                case 2:
                    return food_icons2.length;
                case 3:
                    return food_icons3.length;
                default:
                    return 2;
            }
        }
        @Override
        public Object getItem(int position) {
            switch (TYPE) {
                case 0:
                    return food_icons0[position];
                case 1:
                    return food_icons1[position];
```

```
                    case 2:
                        return food_icons2[position];
                    case 3:
                        return food_icons3[position];
                    default:
                        return 1;
                }
            }
            @Override
            public long getItemId(int position) {
                return position;
            }
            @Override
            public View getView(int position, View convertView, ViewGroup parent) {
                convertView = View.inflate(getActivity(), R.layout.food_item, null);
                TextView foodName = convertView.findViewById(R.id.food_name);
                TextView evaluates = convertView.findViewById(R.id.estimate);
                ImageView food_iv = convertView.findViewById(R.id.food_icon);
                TextView price = convertView.findViewById(R.id.food_price);
                evaluates.setText(food_evaluation[position]);
                price.setText(food_price[position]);
                switch (TYPE) {
                    case 0:
                        food_iv.setBackgroundResource(food_icons0[position]);
                        foodName.setText(food_name0[position]);
                        break;
                    case 1:
                        food_iv.setBackgroundResource(food_icons1[position]);
                        foodName.setText(food_name1[position]);
                        break;
                    case 2:
                        food_iv.setBackgroundResource(food_icons2[position]);
                        foodName.setText(food_name2[position]);
                        break;
                    case 3:
                        food_iv.setBackgroundResource(food_icons3[position]);
                        foodName.setText(food_name3[position]);
                        break;
                    default:
                        food_iv.setBackgroundResource(food_icons0[position]);
                        foodName.setText(food_name0[position]);
                }
                return convertView;
            }
        }
    }
}
```

8. 编写 MainActivity 代码

在 MainActivity 中将 MenuFragment 与 FoodFragment 添加到 MainActivity 界面上，具体代码详见【文件 4-17】。

【文件 4-17】 MainActivity 代码文件

```
public class MainActivity extends AppCompatActivity {
    private FragmentTransaction beginTransaction;
    @Override
```

```java
        protected void onCreate(Bundle savedInstanceState) {
            super.onCreate(savedInstanceState);
            setContentView(R.layout.activity_main);
            // 1. 创建 Fragment 对象
            FoodFragment foodFragment = new FoodFragment();
            MenuFragment menuFragment = new MenuFragment();
            // 2. 获取 FragmentManager 的实例
            FragmentManager fm = getSupportFragmentManager();
            // 3. 开启 FragmentTransaction 事务
            beginTransaction = fm.beginTransaction();
            // 4. 在 Activity 布局中动态添加两个 Fragment(menuFragment 和 foodFragment)
            beginTransaction.replace(R.id.foodcontent, foodFragment);
            beginTransaction.replace(R.id.menucontent, menuFragment);
            // 5. 提交事务
            beginTransaction.commit();
        }
    }
```

上述代码中,通过 replace()方法将 MenuFragment 与 FoodFragment 添加到对应的布局中。

运行上述程序,点击界面左侧分类列表中的"推荐"按钮,可在右侧看到对应的餐品列表信息,实现餐厅点餐功能。

4.6 小结

本章详细地讲解了 Activity 的相关知识,包括 Activity 的生命周期,如何创建、启动和关闭 Activity,Activity 的启动模式和数据传递,以及 Fragment 的使用。在 Android 程序中使用最多的就是 Activity 以及 Activity 之间的数据传递,因此要求读者务必掌握本章内容。

习题 4

一、判断题

1. 如果 Activity 不设置启动模式,则默认为 standard。 （   ）
2. Fragment 与 Activity 的生命周期方法是一致的。 （   ）
3. 如果想要关闭当前的 Activity,可以调用 Activity 提供的 finish()方法。 （   ）
4. <intent-filter>标签中间只能包含一个 action 属性。 （   ）
5. 默认情况下,Activity 的启动方式是 standard。 （   ）

二、填空题

1. Activity 的启动模式包括 standard、singleTop、singleTask 和_____。
2. 启动一个新的 Activity 并且获取这个 Activity 的返回数据,需要重写_____方法。
3. Activity 的_____方法用于关闭当前的 Activity。

4. 打开一个新页面，新页面的生命周期方法依次为 onCreate→_____→_____。
5. 关闭现有的页面，现有页面的生命周期方法依次为 onPause→_____→_____。

三、简答题

1. 简述 Activity 的生命周期的方法以及什么时候被调用。
2. 简述 Activity 的 4 种启动模式及其特点。
3. 简要描述意图 Intent 主要完成哪几项工作。

第 5 章

多媒体应用开发

 **本章导图**

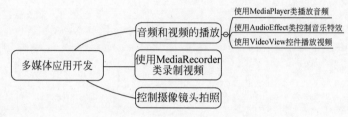

 主要内容

- 使用 MediaPlayer 类播放音频。
- 使用 AudioEffect 类控制音乐特效。
- 使用 VideoView 控件播放视频。
- 使用 MediaRecorder 类录制视频。
- 控制摄像头拍照。

 难点

- 掌握各种音频视频类和控件的使用方法。
- Android 相机拍照功能的实现。

随着手机硬件的不断提升,手机已经成为人们日常生活中必不可少的设备,设备里面的多媒体资源想必是很多人的兴趣所在。多媒体资源一般包括音频、视频等,Android 系统针对不同的多媒体提供了不同的类进行支持。

Android 提供了常见的音频、视频的编码、解码机制,支持的音频格式有 MP3、WAV 和 3GP 等,支持的视频格式有 MP4 和 3GP 等。接下来,本章将针对多媒体应用中的音频和视频操作进行讲解。

5.1 音频和视频的播放

Android 提供了简单的 API 来播放音频、视频,下面将详细介绍如何使用它们。

5.1.1 使用 MediaPlayer 类播放音频

Android 应用中播放音频文件的功能一般都是通过 MediaPlayer 类实现的,该类提供了全面的方法支持多种格式的音频文件。MediaPlayer 类的常用方法如表 5.1 所示。

表 5.1 MediaPlayer 类的常用方法

方　　法	说　　明
setDataSource()	设置要播放音频文件的位置
prepare()	在开始播放之前调用该方法完成准备工作
start()	开始或继续播放音频
pause()	暂停播放音频
reset()	重置 MediaPlayer 对象
seekTo()	从指定位置开始播放音频
stop()	停止播放音频,调用该方法后 MediaPlayer 对象无法再播放音频
release()	释放与 MediaPlayer 对象相关的资源
isPlaying()	判断当前是否正在播放音频
getDuration()	获取载入的音频文件的时长

观看视频

接下来通过演示使用 MediaPlayer 类播放音频的过程来了解 MediaPlayer 的使用,具体如下。

1. 实例化 MediaPlayer 类

使用 MediaPlayer 类播放音频时,首先创建一个 MediaPlayer 类的对象,接着调用 setAudioStreamType() 方法设置音频类型。示例代码如下:

```
MediaPlayer mediaPlayer = new MediaPlayer ();              //创建 MediaPlayer 类的对象
mediaPlayer.setAudioStreamType(AudioManager.STREAM_MUSIC); //设置音频类型
```

上述代码中的 setAudioStreamType() 方法中传递的参数表示音频类型。音频类型有很多种,最常用的音频类型有以下几种。

- AudioManager.STREAM_MUSIC:音乐。
- AudioManager.STREAM_RING:响铃。
- AudioManager.STREAM_ALARM:闹钟。
- AudioManager.STREAM_NOTIFICTION:提示音。

2. 设置数据源

根据音频文件存放的位置不同,将数据源的设置分为三种方式,分别为设置播放应用自带的音频文件、设置播放 SD 卡中的音频文件和设置播放网络音频文件。示例代码如下:

```
//1.设置播放应用自带的音频文件
mediaPlayer = MediaPlayer.create(MainActivity.this, R.raw.xxx);
//2.设置播放 SD 卡中的音频文件
mediaPlayer.setDataSource("SD 卡中的音频文件的路径");
//3.设置播放网络音频文件
mediaPlayer.setDataSource("http://www.xxx.mp3");
```

需要注意的是,播放网络中的音频文件时,需要在清单文件中添加访问网络的权限,示例代码如下。

```
< uses - permission android:name = "android.permission.INTERNET"/>
```

3. 播放音频文件

一般在调用 start( ) 方法播放音频文件之前,程序会调用 prepare( ) 方法或 prepareAsync( )方法将音频文件解析到内存中。prepare( )方法为同步操作,一般用于解析较小的文件;prepareAsync( )方法为异步操作,一般用于解析较大的文件。

(1) 播放小音频文件。

示例代码如下:

```
mediaPlayer.prepare();
mediaPlayer.start();
```

需要注意的是,使用 create( )方法创建 MediaPlayer 对象并设置音频文件时,不能调用 prepare( )方法,直接调用 start( )方法播放音频文件即可。

(2) 播放大音频文件。

示例代码如下:

```
mediaPlayer.prepareAsync();
mediaPlayer.setOnPreparedListener(new OnPreparedlistener){
    public void onPrepared(MediaPlayer player){
        player.start();
    }
}
```

上述代码中,prepareAsync( )方法是子线程中执行的异步操作,不管它是否执行完毕,都不会影响主线程操作。setOnPreparedListener( )方法用于设置 MediaPlayer 类的监听器,用于监听音频文件是否解析完成,如果解析完成,则会调用 onPrepared( )方法,在该方法内部调用 start( )方法播放音频文件。

4. 暂停播放

pause( )方法用于暂停播放音频。在暂停播放之前,首先要判断 MediaPlayer 对象是否存在,并且当前是否正在播放音频。示例代码如下:

```
if(mediaPlayer!= null && mediaPlayer.isPlaying()){
    mediaPlayer.pause();
}
```

5. 重新播放

seekTo( )方法用于定位播放。该方法用于快退或快进音频播放,方法中传递的参数表示将播放时间定在多少毫秒,如果传递的参数为 0,则表示从头开始播放。示例代码如下:

```
//1.播放状态下进行重播
if(mediaPlayer!= null && mediaPlayer.isPlaying()){
    mediaPlayer.seekTo(0);              //设置从头开始播放音频
    return;
}
//2.暂停状态下进行重播,要调用 start()方法
```

```
if(mediaPlayer!= null){
    mediaPlayer.seekTo(0);              //设置从头开始播放音频
    mediaPlayer.start();
}
```

6. 停止播放

stop()方法用于停止播放,停止播放之后还要调用release()方法将MediaPlayer对象占用的资源释放并将该对象设置为null。示例代码如下:

```
if(mediaPlayer!= null && mediaPlayer.isPlaying()){
    mediaPlayer.stop();                 //停止播放
    mediaPlayer.release();              //释放MediaPlayer对象占用的资源
    mediaPlayer = null;
}
```

7. 播放不同来源音频文件的步骤

下面简单归纳一下用MediaPlayer类播放不同来源音频文件的步骤。

1) 播放应用的资源文件

播放应用的资源文件需要如下两步。

(1) 调用MediaPlayer类的create(Context context,int resid)方法加载指定资源文件。

(2) 调用MediaPlayer类的start()、pause()、stop()等方法控制播放。

例如如下代码:

```
MediaPlayer mPlayer = MediaPlayer.create(this,R.raw.song);
mPlayer.start();
```

提示:音频资源文件一般放在Android应用的/res/raw目录下。

2) 播放应用的原始资源文件

播放应用的原始资源文件按如下步骤执行。

(1) 调用Context对象的getAssets()方法获取应用的AssetManager。

(2) 调用AssetManager对象的openFd(String name)方法打开指定的原始资源,该方法返回一个AssetFileDescriptor对象。

(3) 调用AssetFileDescriptor对象的getFileDescriptor()、getStartOffset()和getLength()方法来获取音频文件的文件描述符、开始位置、长度等。

(4) 创建MediaPlayer对象,并调用MediaPlayer对象的setDataSource(FileDescriptor fd,long offset,long length)方法来装载音频资源。

(5) 调用MediaPlayer对象的prepare()方法准备音频。

(6) 调用MediaPlayer对象的start()、pause()、stop()等方法控制播放。

注意,虽然MediaPlayer对象提供了setDataSource(FileDescriptor fd)方法来装载指定音频资源,但实际使用时这个方法似乎有问题:不管程序调用openFd(String name)方法时指定打开哪个原始资源,MediaPlayer将总是播放第一个原始的音频资源。

例如如下代码片段:

```
AssetManager am = getAssets();
//打开指定音乐文件
AssetFileDescriptor afd = am.openFd(music);
MediaPlayer mPlayer = new MediaPlayer();
//使用 Mediaplayer 加载指定的声音文件
mPlayer.setDataSource(afd.getFileDescriptor()
    , afd.getStartOffset()
    , afd.getLength());
//准备声音
mPlayer.prepare();
//播放
mPlayer.start();
```

3）播放外部存储器上的音频文件

播放外部存储器上的音频文件按如下步骤执行。

（1）创建 MediaPlayer 对象，并调用 MediaPlayer 对象的 setDateSource(String path)方法装载指定的音频文件。

（2）调用 MediaPlayer 对象的 prepare()方法准备音频。

（3）调用 MediaPlayer 对象的 start()、pause()、stop()等方法控制播放。

例如如下代码：

```
MediaPlayer mPlayer = new MediaPlayer();
//使用 MediaPlayer 加载指定的声音文件
mPlayer.setDataSource("/mnt/sdcard/mysong.mp3");
//准备声音
mPlayer.prepare();
//播放
mPlayer.start();
```

4）播放来自网络的音频文件

播放来自网络的音频文件有两种方式：①直接使用 MediaPlayer 对象的静态 create (Contextcontext，Uri uri)方法；②调用 MediaPlayer 对象的 setDataSource（Context context，Uri uri)装载指定 Uri 对应的音频文件。

以第二种方式播放来自网络的音频文件的步骤如下。

（1）根据网络上的音频文件所在的位置创建 Uri 对象。

（2）创建 MediaPlayer 对象，并调用 MediaPlayer 对象的 setDateSource（Context context，Uri uri)方法装载 Uri 对应的音频文件。

（3）调用 MediaPlayer 对象的 prepare()方法准备音频。

（4）调用 MediaPlayer 对象的 start()、pause()、stop()等方法控制播放。

例如如下代码片段：

```
Uri uri = Uri.parse("http://www.crazyit.org/abc.mp3");
MediaPlayer mPlayer = new MediaPlayer();
//使用 MediaPlayer 根据 Uri 来加载指定的声音文件
mPlayer.setDataSource(this, uri);
//准备声音
mPlayer.prepare();
```

```
//播放
mPlayer.start();
```

MediaPlayer 除了调用 prepare() 方法来准备声音之外,还可以调用 prepareAsync() 方法来准备声音。prepareAsync() 方法与普通 prepare() 方法的区别在于,prepareAsync() 方法是异步的,它不会阻塞当前的 UI 线程。

归纳起来,MediaPlayer 的状态图如图 5.1 所示。

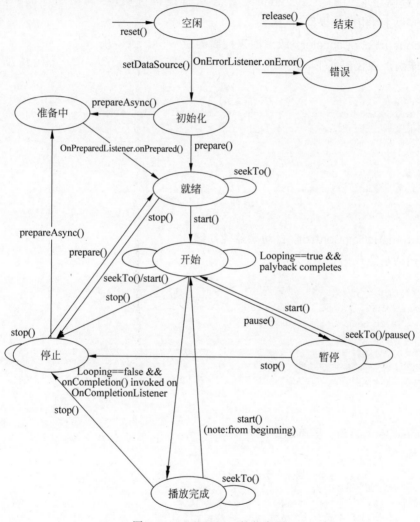

图 5.1　MediaPlayer 的状态图

5.1.2　使用 AudioEffect 类控制音乐特效

Android 可以控制播放音乐时的均衡器、重低音、音场及显示音乐波形等,这些都是靠 AudioEffect 及其子类来完成的,它包含如下常用子类:

- AcousticEchoCanceler：取消回声控制器。
- AutomaticGainControl：自动增益控制器。

- NoiseSppressor：噪声压制控制器。
- BassBoost：重低音控制器。
- Equalizer：均衡控制器。
- PresetReverb：预设音场控制器。
- Visualizer：示波器。

上面的子类中前三个子类的用法很简单，只要调用它们的静态 create()方法创建相应的实例，然后调用它们的 isAvailable()方法判断是否可用，再调用 setEnabled(boolean enabled)方法启用相应效果即可。

1. AcousticEchoCanceler：取消回声控制器

该功能的示意代码如下：

```
//获取取消回声控制器
AcousticEchoCanceler canceler = AcousticEchoCanceler.create(0
    , mPlayer.getAudioSessionId())
if(canceler.isAvailable())
{
    //启用取消回声功能
    canceler.setEnabled(true);
}
```

2. AutomaticGainControl：自动增益控制器

该功能的示意代码如下：

```
//获取自动增益控制器
AutomaticGainControl ctrl = AutomaticGainControl.create(0
    , mPlayer.getAudioSessionId())
if(ctrl.isAvailable())
{
    //启用自动增益控制功能
    ctrl.setEnabled(true);
}
```

3. NoiseSppressor：噪声压制控制器

该功能的示意代码如下：

```
//获取噪声压制控制器
NoiseSuppressor suppressor = NoiseSuppressor.create(0
    , mPlayer.getAudioSessionId())
if(suppressor.isAvailable())
{
    //启用噪声压制功能
    suppressor.setEnabled(true);
}
```

BassBoost、Equalizer、PresetReverb、Visualizer 这 4 个类，都需要调用构造器来创建实例。创建实例时，同样需要传入一个 audioSession 参数，为了启用它们，同样需要调用 AudioEffect 基类的 setEnabled(true)方法。

4. BassBoost：重低音控制器

低音增强,用于增强或放大声音的低频。它与简单的均衡器相当,但仅限于低频范围内的一个频段放大。

获取 BassBoost 对象之后,可调用它的 setStrength(short strength)方法来设置重低音的强度。示例代码如下:

```
BassBoost bassBoost = new BassBoost(0,mediaPlayer.getAudioSessionId());
bassBoost.setEnabled(true);
if (bassBoost.getStrengthSupported()){
    bassBoost.setStrength((short) 100);
}
```

其中,getStrengthSupported()表示是否支持设置强度。如果此方法返回 false,则仅支持一种强度,并且 setStrength() 方法始终舍入到该值。

setStrength()设置效果的当前强度,强度的有效范围是[0,1000],0 表示最温和的效果,1000 表示最强的效果。

5. Equalizer：均衡控制器

Equalizer 提供了 getNumberOfPresets()方法获取系统所有预设的音场,并提供了 getPresetName()方法获取预设音场名称。获取 Equalizer 对象之后,可调用它的 getNumberOfBands()方法获取该均衡器支持的总频率数,再调用 getCenterFreq(short band)方法根据索引来获取频率。当用户想为某个频率的均衡器设置参数值时,可调用 setBandLevel(short band, short level)方法进行设置。示例代码如下:

```
MediaPlayer mediaPlayer = MediaPlayer.create(this, R.raw.test_cbr/*音频路径*/);
Equalizer equalizer = new Equalizer(0, mediaPlayer.getAudioSessionId());
equalizer.setEnabled(true);
//获取均衡器引擎支持的频段数
short bands = equalizer.getNumberOfBands();
//获取最大和最小增益
final short minEQLevel = equalizer.getBandLevelRange()[0];
final short maxEQLevel = equalizer.getBandLevelRange()[1];
for (short i = 0; i < bands; i++) {
    final short band = i;
    //获取当前频段的中心频率,分别为 60Hz,230Hz,910Hz,3600Hz,14000Hz
    int currentFreq = equalizer.getCenterFreq(band);
    //获取给定均衡器频段的增益
    short level = equalizer.getBandLevel(band);
    Log.d("Equalizer","currentFreq is: " + currentFreq + ", band level is: " + level);
    //为给定的均衡器频带设置增益值
    equalizer.setBandLevel(band,xx);
}
```

在构造函数 Equalizer(int priority, int audioSession)中,参数如下:

int priority：优先级,多个应用可以共享同一 Equalizer 引擎,该参数指出控制优先权,默认为 0。

int audioSession：音频会话 ID,系统范围内唯一,Equalizer 将被附加在拥有相同音频会话 ID 的 MediaPlayer 或 AudioTrack 上生效。

Android 系统预置了一些增益参数,可通过下面代码获取:

```
short presets = equalizer.getNumberOfPresets();
//获取系统预设的增益
for (short i = 0; i < presets; i++) {
    Log.d("presets",equalizer.getPresetName(i));
}
```

结果为 Normal、Classical、Dance、Flat、Folk、Heavy Metal、Hip hop、Jazz、Pop、Rock。
然后通过 equalizer.usePreset();使用系统预置参数。

销毁时:

```
if (equalizer != null) {
    equalizer.setEnabled(false);
    equalizer.release();
    equalizer = null;
}
```

6. PresetReverb:预设音场控制器

PresetReverb 使用预设混响来配置全局混响,适合于音乐。预置的常见混响场景有:
PresetReverb.PRESET_LARGEHALL:适合整个管弦乐队的大型大厅。
PresetReverb.PRESET_LARGEROOM:适合现场表演的大型房间的混响预设。
获取 PresetReverb 对象之后,可调用它的 setPreset(short preset)方法设置使用预设置的音场。示例代码如下:

```
PresetReverb presetReverb = new PresetReverb(0,mediaPlayer.getAudioSessionId());
presetReverb.setEnabled(true);
presetReverb.setPreset(PresetReverb.PRESET_LARGEROOM);
```

7. Visualizer:示波器

Visualizer 对象并不用于控制音乐播放效果,它只是显示音乐的播放波形。为了实时显示该示波器的数据,需要为该组件设置一个 OnDataCaptureListener 监听器,该监听器将负责更新波形显示组件的界面。

5.1.3 使用 VideoView 控件播放视频

播放视频与播放音频相比,播放视频需要使用视觉控件将影像展示出来。Android 系统中的 VideoView 控件就是播放视频用的,借助它可以完成一个简易的视频播放器。
VideoView 控件提供了一些用于控制视频播放的方法,如表 5.2 所示。

表 5.2 VideoView 控件的常用方法

方 法	说 明
setVideoPath()	设置要播放的视频文件的位置
start()	开始或继续播放视频
pause()	暂停播放视频
resume()	重新开始播放视频

续表

方　法	说　明
seekTo()	从指定位置开始播放视频
isPlaying()	判断当前是否正在播放视频
getDuration()	获取载入的视频文件的时长

接下来讲解如何通过 VideoView 控件播放视频的过程，具体介绍如下。

1. 在布局文件中添加 VideoView 控件

如果想在界面上播放视频，则首先需要在布局文件中放置 1 个 VideoView 控件用于显示视频播放界面。在布局中添加 VideoView 控件的示例代码如下：

```
<VideoView
    android:id = "@ + id/videoview"
    android:layout_width = "match_parent"
    android:layout_height = "match_parent" />
```

2. 视频的播放

使用 VideoView 控件既可以播放本地存放的视频，也可以播放网络中的视频。示例代码如下：

```
VideoView videoView = (VideoView) findViewById(R.id.videoview);
videoView.setVideoPath("mnt/sdcard/xxx.avi");                    //播放本地视频
videoView.setVideoURI(Uri.parse("http://www.xxx.avi"));          //加载网络视频
videoView.start();                                               //播放视频
```

根据上述代码可知，播放本地视频时需要调用 VideoView 控件的 setVideoPath()方法，将本地视频地址传入该方法中即可。播放网络视频时需要调用 VideoView 控件的 setVideoURI()方法，通过调用 parse()方法将网络视频地址转换为 Uri 并传递到 setVideoURI()方法中。

需要注意的是，播放网络视频时需要在 AndroidManifest.xml 文件的<manifest>标签中添加访问网络的权限。示例代码如下：

```
<uses - permission android:name = "android.permission.INTERNET"/>
```

3. 为 VideoView 控件添加控制器

使用 VideoView 控件播放视频时，可以通过 setMediaController()方法为它添加一个控制器 MediaController，该控制器中包含媒体播放器（MediaPlayer）中的一些典型按钮，如播放/暂停（Play/Pause）、倒带（Rewind）、快进（Fast Forward）与进度滑动器（Progress Slider）等。VideoView 控件能够绑定媒体播放器，从而使播放状态和控件中显示的图像同步。示例代码如下：

```
MediaController controller = new MediaController(context);
videoView.setMediaController(controller);                  //为 VideoView 控件绑定控制器
```

具体的使用详见本章实例视频播放器的实现。

5.2 使用 MediaRecorder 类录制音频

手机一般都提供了麦克风硬件,而 Android 系统就可以利用该硬件来录制音频了。

为了在 Android 应用中录制音频,Android 提供了 MediaRecorder 类。使用 MediaRecorder 类录制音频的过程很简单,按如下步骤进行即可。

(1) 创建 MediaRecorder 对象。

(2) 调用 MediaRecorder 对象的 setAudioSource() 方法设置声音来源,一般传入 MediaRecorder.AudioSource.MIC 参数指定录制来自麦克风的声音。

(3) 调用 MediaRecorder 对象的 setOutputFormat() 方法设置所录制的音频文件格式。

(4) 调用 MediaRecorder 对象的 setAudioEncoder()、setAudioEncodingBitRate(int bitrate)、setAudioSamplingRate(int samplingRate)方法设置所录制的声音编码格式、编码位率、采样率等,这些参数将可以控制所录制的声音品质、文件大小。一般来说,声音品质越好,声音文件越大。

(5) 调用 MediaRecorder 对象的 setOutputFile(String path)方法设置录制的音频文件的保存位置。

(6) 调用 MediaRecorder 对象的 prepare()方法准备录制。

(7) 调用 MediaRecorder 对象的 start()方法开始录制。

(8) 录制完成,调用 MediaRecorder 对象的 stop()方法停止录制,并调用 release()方法释放资源。

注意,上面的步骤中第(3)、(4)步千万不能搞反,否则程序将会抛出 IllegalStateException 异常。

下面的程序示范了如何使用 MediaRecorder 类来录制声音,该程序的界面布局很简单,只提供了两个简单的按钮来控制录音开始、停止,故此处不再给出界面布局文件。程序代码如下:

```java
public class RecordSound extends Activity
    implements OnClickListener
{
    //程序中的两个按钮
    ImageButton record , stop;
    //系统的音频文件
    File soundFile ;
    MediaRecorder mRecorder;
    @Override
    public void onCreate(Bundle savedInstanceState)
    {
        super.onCreate(savedInstanceState);
        setContentView(R.layout.main);
        //获取程序界面中的两个按钮
        record = (ImageButton) findViewById(R.id.record);
        stop = (ImageButton) findViewById(R.id.stop);
```

```java
        //为两个按钮的点击事件绑定监听器
        record.setOnClickListener(this);
        stop.setOnClickListener(this);
    }
    @Override
    public void onDestroy()
    {
        if (soundFile != null && soundFile.exists())
        {
            //停止录音
            mRecorder.stop();
            //释放资源
            mRecorder.release();
            mRecorder = null;
        }
        super.onDestroy();
    }
    @Override
    public void onClick(View source)
    {
        switch (source.getId())
        {
            //点击 record 按钮
            case R.id.record:
                if (!Environment.getExternalStorageState().equals(
                    android.os.Environment.MEDIA_MOUNTED))
                {
                    Toast.makeText(RecordSound.this
                        , "SD 卡不存在,请插入 SD 卡!"
                        , 5000)
                        .show();
                    return;
                }
                try
                {
                    //创建保存录音的音频文件
                    soundFile = new File(Environment
                        .getExternalStorageDirectory()
                        .getCanonicalFile() + "/sound.amr");
                    mRecorder = new MediaRecorder();
                    //设置录音的声音来源
                    mRecorder.setAudioSource(MediaRecorder.AudioSource.MIC);
                    //设置录制的声音的输出格式(必须在设置声音编码格式之前设置)
                    mRecorder.setOutputFormat(MediaRecorder
                        .OutputFormat.THREE_GPP);
                    //设置声音编码的格式
                    mRecorder.setAudioEncoder(MediaRecorder
                        .AudioEncoder.AMR_NB);
                    mRecorder.setOutputFile(soundFile.getAbsolutePath());
                    mRecorder.prepare();
                    //开始录音
                    mRecorder.start();                    //①
                }
```

```
                catch (Exception e)
                {
                    e.printStackTrace();
                }
                break;
            //点击 stop 按钮
            case R.id.stop:
                if (soundFile != null && soundFile.exists())
                {
                    //停止录音
                    mRecorder.stop();           //②
                    //释放资源
                    mRecorder.release();        //③
                    mRecorder = null;
                }
                break;
        }
    }
}
```

上面的程序中大段的粗体字代码用于设置录音的相关参数,例如输出文件的格式、声音来源等。上面的程序中①粗体字代码控制 MediaRecorder 类开始录音;当用户点击 stop 按钮时,程序中②号代码控制 MediaRecorder 类停止录音,③号粗体字代码用于释放资源。

录音完成后将可以看到/mnt/sdcard/目录下生成一个 sound.amr 文件,这就是录制的音频文件。Android 模拟器将会直接使用宿主机上的麦克风,因此如果读者的手机上有麦克风,那么该程序即可正常录制声音。

上面的程序需要使用系统的麦克风进行录音,因此需要向该程序授予录音的权限,也就是在 AndroidManifest.xml 文件中增加如下配置:

```
<!-- 授予该程序录制声音的权限 -->
<uses-permission android:name="android.permission.RECORD_AUDIO"/>
```

5.3 控制摄像头拍照

现在的手机一般都会提供相机功能,有些相机的镜头甚至支持 1000 万以上像素,有些甚至支持光学变焦,这些手机已经变成了专业数码相机。为了充分利用手机上的相机功能,Android 应用可以控制拍照和录制视频。

5.3.1 通过 Camera 进行拍照

Android 应用提供了 Camera 来控制拍照,使用 Camera 进行拍照也比较简单,按如下步骤进行即可。

(1) 调用 Camera 的 open()方法打开相机。

(2) 调用 Camera 的 getParameters()方法获取拍照参数。该方法返回一个 Camera.Parameters 对象。

（3）调用 Camera.Parameters 对象方法设置相机参数。

（4）调用 Camera 的 setParameters()方法，并将 Camera.Parameters 对象作为参数传入，这样即可对相机的拍照参数进行控制。

（5）调用 Camera 的 startPreview()方法开始预览取景，在预览取景之前需要调用 Camera 的 setPreviewDisplay(SurfaceHolder holder)方法设置使用哪个 SurfaceView 来显示取景图片。

（6）调用 Camera 的 takePicture()方法进行拍照。

（7）结束程序时，调用 Camera 的 stopPreview()方法结束取景预览，并调用 release()方法释放资源。

下面的程序示范了使用 Camera 来进行拍照，该程序的界面中只提供了一个 SurfaceView 组件来显示预览取景，十分简单。程序代码如下：

```java
public class CaptureImage extends Activity
{
    SurfaceView sView;
    SurfaceHolder surfaceHolder;
    int screenWidth, screenHeight;
    //定义系统所用的照相机
    Camera camera;
    //是否在浏览中
    boolean isPreview = false;
    @Override
    public void onCreate(Bundle savedInstanceState)
    {
        super.onCreate(savedInstanceState);
        //设置全屏
        requestWindowFeature(Window.FEATURE_NO_TITLE);
        getWindow().setFlags(WindowManager.LayoutParams.FLAG_FULLSCREEN,
            WindowManager.LayoutParams.FLAG_FULLSCREEN);
        setContentView(R.layout.main);
        WindowManager wm = (WindowManager) getSystemService(
            Context.WINDOW_SERVICE);
        Display display = wm.getDefaultDisplay();
        //获取屏幕的宽和高
        screenWidth = display.getWidth();
        screenHeight = display.getHeight();
        //获取界面中 SurfaceView 组件
        sView = (SurfaceView) findViewById(R.id.sView);
        //获得 SurfaceView 的 SurfaceHolder
        surfaceHolder = sView.getHolder();
        //为 surfaceHolder 添加一个回调监听器
        surfaceHolder.addCallback(new Callback()
        {
            @Override
            public void surfaceChanged(SurfaceHolder holder, int format, int width,
                int height)
            {
            }
            @Override
```

```java
            public void surfaceCreated(SurfaceHolder holder)
            {
                //打开摄像头
                initCamera();
            }
            @Override
            public void surfaceDestroyed(SurfaceHolder holder)
            {
                //如果 camera 不为 null,则释放摄像头
                if (camera != null)
                {
                    if (isPreview)
                        camera.stopPreview();
                    camera.release();
                    camera = null;
                }
            }
        });
        //设置该 SurfaceView 自己不维护缓冲
        surfaceHolder.setType(SurfaceHolder.SURFACE_TYPE_PUSH_BUFFERS);
    }

    private void initCamera()
    {
        if (!isPreview)
        {
            camera = Camera.open();
        }
        if (camera != null && !isPreview)
        {
            try
            {
                Camera.Parameters parameters = camera.getParameters();
                //设置预览照片的大小
                parameters.setPreviewSize(screenWidth, screenHeight);
                //每秒显示 4 帧
                parameters.setPreviewFrameRate(4);
                //设置图片格式
                parameters.setPictureFormat(PixelFormat.JPEG);
                //设置 JPG 照片的质量
                parameters.set("jpeg-quality", 85);
                //设置照片的大小
                parameters.setPictureSize(screenWidth, screenHeight);
                camera.setParameters(parameters);
                //通过 SurfaceView 显示取景画面
                camera.setPreviewDisplay(surfaceHolder);         //①
                //开始预览
                camera.startPreview();                           //②
                //自动对焦
                camera.autoFocus(null);
            }
            catch (Exception e)
            {
```

```java
            e.printStackTrace();
        }
        isPreview = true;
    }
}

@Override
public boolean onKeyDown(int keyCode, KeyEvent event)
{
    switch (keyCode)
    {
        //当用户按照相键、中央键时执行拍照
        case KeyEvent.KEYCODE_DPAD_CENTER:
        case KeyEvent.KEYCODE_CAMERA:
            if (camera != null && event.getRepeatCount() == 0)
            {
                //拍照
                camera.takePicture(null, null , myjpegCallback);//③
                return true;
            }
            break;
    }
    return super.onKeyDown(keyCode, event);
}

PictureCallback myjpegCallback = new PictureCallback()
{
    @Override
    public void onPictureTaken(byte[] data, Camera camera)
    {
        //根据拍照所得的数据创建位图
        final Bitmap bm = BitmapFactory.decodeByteArray(data
            , 0, data.length);
        //加载/layout/save.xml 文件对应的布局资源
        View saveDialog = getLayoutInflater().inflate(
            R.layout.save, null);
        final EditText photoName = (EditText) saveDialog
            .findViewById(R.id.phone_name);
        //获取 saveDialog 对话框上的 ImageView 控件
        ImageView show = (ImageView) saveDialog.findViewById(R.id.show);
        //显示刚刚拍得的照片
        show.setImageBitmap(bm);
        //使用对话框显示 saveDialog 控件
        new AlertDialog.Builder(CaptureImage.this)
            .setView(saveDialog)
            .setPositiveButton("保存", new OnClickListener()
            {
                @Override
                public void onClick(DialogInterface dialog,
                    int which)
                {
                    //创建一个位于 SD 卡上的文件
                    File file = new File(Environment.getExternalStorageDirectory()
```

```
                            , photoName.getText().toString() + ".jpg");
                    FileOutputStream outStream = null;
                    try
                    {
                        //打开指定文件对应的输出流
                        outStream = new FileOutputStream(file);
                        //把位图输出到指定文件中
                        bm.compress(CompressFormat.JPEG, 100, outStream);
                        outStream.close();
                    }
                    catch (IOException e)
                    {
                        e.printStackTrace();
                    }
                }
            })
            .setNegativeButton("取消", null)
            .show();
        //重新浏览
        camera.stopPreview();
        camera.startPreview();
        isPreview = true;
    }
};
}
```

上面的程序中大段粗体字代码用于设置相机的拍照参数,这些参数可以控制图片的品质和图片文件的大小。

上面的程序中①号粗体字代码设置使用指定的 SurfaceView 来显示取景预览图片,程序中②号粗体字代码则用于开始预览取景。当用户按下相机的拍照键或中央键时,程序的③号粗体字代码调用 takePicture() 方法进行拍照。

调用 takePicture() 方法进行拍照时传入了一个 PictureCallback 对象——当程序获取了拍照所得的图片数据之后,PictureCallback 对象将会被回调,该对象负责保存图片或将其上传到网络。

Android 模拟器不会使用宿主机上的摄像头作为相机镜头,因此取景预览将总是一片空白,这是因为模拟器没有摄像头的缘故。当用户进行拍照时,系统将会使用一张已有的图片作为图片。

用户在对话框中输入图片的名称后,程序将会把拍得的图片保存到 SD 卡上。

运行该程序需要获得相机拍照的权限,因此需要在 AndroidManifest.xml 文件中增加如下代码片段进行授权:

```xml
<!-- 授予程序使用摄像头的权限 -->
<uses-permission android:name="android.permission.CAMERA"/>
<uses-feature android:name="android.hardware.camera"/>
<uses-feature android:name="android.hardware.camera.autofocus"/>
```

5.3.2 录制视频短片

MediaRecorder 类除了可用于录制音频之外,还可用于录制视频。使用 MediaRecorder

类录制视频与录制音频的步骤基本相同。只是录制视频时不仅需要采集声音,还需要采集图像。为了让 MediaRecorder 类录制时采集图像,应该在调用 setAudioSource(int audio_source)方法时再调用 setVideoSource(int video_source)方法来设置图像来源。

除此之外,还需在调用 setOutputFormat()方法设置输出文件格式之后进行如下步骤。

(1) 调用 MediaRecorder 对象的 setVideoEncoder()、setVideoEncodingBitRate(int bitRate)、setVideoFrameRate()方法设置所录制的视频的编码格式、编码位率、每秒多少帧等,这些参数将可以控制所录制的视频的品质、文件的大小。一般来说,视频品质越好,视频文件越大。

(2) 调用 MediaRecorder 对象的 setPreviewDisplay(Surface sv)方法设置使用哪个 SurfaceView 控件来显示视频预览。

剩下的代码则与录制音频的代码基本相同。下面的程序示范了如何录制视频,该程序的界面中提供了两个按钮用于控制开始和结束录制;除此之外,程序界面中还提供了一个 SurfaceView 控件来显示视频预览。该程序的界面布局文件如下:

```xml
<?xml version = "1.0" encoding = "utf-8"?>
<LinearLayout
    xmlns:android = "http://schemas.android.com/apk/res/android"
    android:orientation = "vertical"
    android:layout_width = "fill_parent"
    android:layout_height = "fill_parent"
    android:gravity = "center_horizontal"
    >
<LinearLayout
    android:orientation = "horizontal"
    android:layout_width = "wrap_content"
    android:layout_height = "wrap_content"
    android:gravity = "center_horizontal"
    >
<ImageButton
    android:id = "@+id/record"
    android:layout_width = "wrap_content"
    android:layout_height = "wrap_content"
    android:src = "@drawable/record"
    />
<ImageButton
    android:id = "@+id/stop"
    android:layout_width = "wrap_content"
    android:layout_height = "wrap_content"
    android:src = "@drawable/stop"
    />
</LinearLayout>
<!-- 显示视频预览的 SurfaceView -->
<SurfaceView
    android:id = "@+id/sView"
    android:layout_width = "fill_parent"
    android:layout_height = "fill_parent"
    />
</LinearLayout>
```

提供了上面所示的界面布局文件之后，接下来就可以在程序中使用 MediaRecorder 类来录制视频了。录制视频与录制音频的步骤基本相似，只是需要额外设置视频的图像来源、视频格式等，除此之外还需要设置使用 SurfaceView 控件显示视频预览。录制视频的程序代码如下：

```java
public class RecordVideo extends Activity
    implements OnClickListener
{
    //程序中的两个按钮
    ImageButton record , stop;
    //系统的视频文件
    File videoFile ;
    MediaRecorder mRecorder;
    //显示视频预览的 SurfaceView
    SurfaceView sView;
    //记录是否正在进行录制
    private boolean isRecording = false;

    @Override
    public void onCreate(Bundle savedInstanceState)
    {
        super.onCreate(savedInstanceState);
        setContentView(R.layout.main);
        //获取程序界面中的两个按钮
        record = (ImageButton) findViewById(R.id.record);
        stop = (ImageButton) findViewById(R.id.stop);
        //让 stop 按钮不可用
        stop.setEnabled(false);
        //为两个按钮的单击事件绑定监听器
        record.setOnClickListener(this);
        stop.setOnClickListener(this);
        //获取程序界面中的 SurfaceView 控件
        sView = (SurfaceView) this.findViewById(R.id.sView);
        //下面设置 Surface 不需要自己维护缓冲区
        sView.getHolder().setType(SurfaceHolder.SURFACE_TYPE_PUSH_BUFFERS);
        //设置分辨率
        sView.getHolder().setFixedSize(320, 280);
        //设置该组件让屏幕不会自动关闭
        sView.getHolder().setKeepScreenOn(true);
    }

    @Override
    public void onClick(View source)
    {
        switch (source.getId())
        {
            //点击 record 按钮
            case R.id.record:
                if (!Environment.getExternalStorageState().equals(
                    android.os.Environment.MEDIA_MOUNTED))
                {
                    Toast.makeText(RecordVideo.this
```

```
            ,"SD 卡不存在,请插入 SD 卡!"
            , 5000)
            .show();
        return;
    }
    try
    {
        //创建保存录制视频的视频文件
        videoFile = new File(Environment
            .getExternalStorageDirectory()
            .getCanonicalFile() + "/myvideo.mp4");
        //创建 MediaPlayer 对象
        mRecorder = new MediaRecorder();
        mRecorder.reset();
        //设置从麦克风采集声音
        mRecorder.setAudioSource(MediaRecorder.AudioSource.MIC);
        //设置从摄像头采集图像
        mRecorder.setVideoSource(MediaRecorder.VideoSource.CAMERA);
        //设置视频文件的输出格式(必须在设置声音编码格式、
        //图像编码格式之前设置)
        mRecorder.setOutputFormat(MediaRecorder
            .OutputFormat.MPEG_4);
        //设置声音编码的格式
        mRecorder.setAudioEncoder(MediaRecorder
            .AudioEncoder.DEFAULT);
        //设置图像编码的格式
        mRecorder.setVideoEncoder(MediaRecorder
            .VideoEncoder.MPEG_4_SP);
        mRecorder.setVideoSize(320, 280);
        //每秒 4 帧
        mRecorder.setVideoFrameRate(4);
        mRecorder.setOutputFile(videoFile.getAbsolutePath());
        //指定使用 SurfaceView 来预览视频
        mRecorder.setPreviewDisplay(sView.getHolder().get Surface());   //①
        mRecorder.prepare();
        //开始录制
        mRecorder.start();
        System.out.println("--- recording ---");
        //让 record 按钮不可用
        record.setEnabled(false);
        //让 stop 按钮可用
        stop.setEnabled(true);
        isRecording = true;
    }
    catch (Exception e)
    {
        e.printStackTrace();
    }
    break;
//点击 stop 按钮
case R.id.stop:
    //如果正在进行录制
    if (isRecording)
```

```
                {
                    //停止录制
                    mRecorder.stop();
                    //释放资源
                    mRecorder.release();
                    mRecorder = null;
                    //让 record 按钮可用
                    record.setEnabled(true);
                    //让 stop 按钮不可用
                    stop.setEnabled(false);
                }
                break;
        }
    }
}
```

上面的程序中粗体字代码设置了视频所采集的图像来源,以及视频的压缩格式、视频分辨率等属性,程序的①号粗体字代码则用于设置使用 SurfaceView 控件显示指定视频预览。

运行该程序需要使用麦克风录制声音,需要使用摄像头采集图像,这些都需要授予相应的权限。不仅如此,由于该录制视频时视频文件增大得较快,可能需要使用外部存储器,因此需要对应用程序授予相应的权限。也就是需要在 AndroidManifest.xml 文件中增加如下授权配置:

```
<!-- 授予该程序录制声音的权限 -->
<uses-permission android:name="android.permission,RECORD_AUDIO"/>
<!-- 授予该程序使用摄像头的权限 -->
<uses-permission android:name="android.permission.CAMERA"/>
<uses-permission android:name="android.permission,MOUNT_UNMOUNT_FILESYSTEMS"/>
<!-- 授予使用外部存储器的权限 -->
<uses-permission android:name="android.permission.WRITE_EXTERNAL_STORAGE"/>
```

当在模拟器上运行该程序时,由于模拟器上没有摄像头硬件支持,因此程序无法采集视频所需的图像,所以建议在有摄像头硬件支持的真机上运行该程序。

5.4 应用实例:视频播放器

本章前面讲解了如何通过 VideoView 控件播放视频的相关知识,接下来我们以图 5.2 所示的界面为例,讲解如何通过 VideoView 控件实现一个视频播放器的案例,具体步骤如下:

1. 创建程序

创建一个名为 VideoView 的应用程序,包名指定为 cn.itcast.videoview。

2. 导入视频文件

选中 res 文件夹,在该文件夹中创建一个 raw 文件夹,将视频文件 video.mp4 放入 raw 文件夹中,如图 5.3 所示。

3. 放置界面控件

在 activity_main.xml 文件中,放置 1 个 ImageView 控件用于显示播放(暂停)按钮,1 个 VideoView 控件用于显示视频。完整布局代码如下:

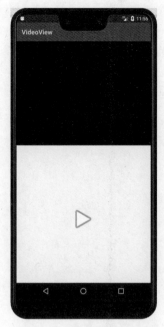

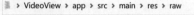

video.mp4

图 5.2　视频播放器界面　　　　　　　图 5.3　raw 文件夹

```xml
<?xml version = "1.0" encoding = "utf-8"?>
<RelativeLayout xmlns:android = "http://schemas.android.com/apk/res/android"
    xmlns:tools = "http://schemas.android.com/tools"
    android:layout_width = "match_parent"
    android:layout_height = "match_parent"
    tools:context = ".MainActivity">
    <ImageView
        android:id = "@+id/bt_play"
        android:layout_width = "80dp"
        android:layout_height = "80dp"
        android:layout_alignParentBottom = "true"
        android:layout_centerHorizontal = "true"
        android:layout_marginBottom = "150dp"
        android:src = "@android:drawable/ic_media_play" />
    <VideoView
        android:id = "@+id/videoview"
        android:layout_width = "match_parent"
        android:layout_height = "match_parent" />
</RelativeLayout>
```

4. 编写界面交互代码

在 MainActivity 中创建一个 play()方法，在该方法中实现视频播放的功能。具体代码如下：

```java
public class MainActivity extends AppCompatActivity implements
        View.OnClickListener {
    private VideoView videoView;
    private MediaController controller;
```

```
        ImageView iv_play;
        @Override
        protected void onCreate(Bundle savedInstanceState) {
            super.onCreate(savedInstanceState);
            setContentView(R.layout.activity_main);
            videoView = (VideoView) findViewById(R.id.videoview);
            iv_play = (ImageView) findViewById(R.id.bt_play);
            //拼出在资源文件夹下的视频文件路径字符串
            String url = "android.resource://" + getPackageName() + "/" + R.raw.video;
            //字符串解析成Uri
            Uri uri = Uri.parse(url);
            //设置VideoView的播放资源
            videoView.setVideoURI(uri);
            //VideoView绑定控制器
            controller = new MediaController(this);
            videoView.setMediaController(controller);
            iv_play.setOnClickListener(this);
        }
        @Override
        public void onClick(View v) {
            switch (v.getId()) {
                case R.id.bt_play:
                    play();
                    break;
            }
        }
        //播放视频
        private void play() {
            if (videoView != null && videoView.isPlaying()) {
                iv_play.setImageResource(android.R.drawable.ic_media_play);
                videoView.stopPlayback();
                return;
            }
            videoView.start();
            iv_play.setImageResource(android.R.drawable.ic_media_pause);
            videoView.setOnCompletionListener(new MediaPlayer.OnCompletionListener() {
                @Override
                public void onCompletion(MediaPlayer mp) {
                    iv_play.setImageResource(android.R.drawable.ic_media_play);
                }
            });
        }
```

上述代码中，通过setVideoURI()方法将视频文件的路径加载到VideoView控件上，并通过setMediaController()方法为VideoView控件绑定控制器，该控制器可以显示视频的播放、暂停、快进、快退和进度条等按钮。

重写了onClick()方法，在该方法中调用play()方法用于播放视频。

创建了一个play()方法，在该方法中首先通过isPlaying()方法判断当前视频是否正在播放，如果正在播放，则通过setImageResource()方法设置播放按钮的图片为ic_media_play.png，接着调用stopPlayback()方法停止播放视频，否则，调用start()方法播放视频，并设置暂停按钮的图片为 ic_media_pause.png，接着通过 setOnCompletionListener()方法设置 VideoView 控件的监听器，当视频播放完时，会调用该监听器中的 onCompletion()方法，在该方法中将播放按钮的图片设置为 ic_media_play.png。

运行结果如图 5.4 和图 5.5 所示。

图 5.4　视频播放器运行结果 1　　　图 5.5　视频播放器运行结果 2

5.5　小结

本章主要讲解了音频、视频的播放过程以及音频的录制过程，同时介绍了使用手机的摄像头进行视频的录制过程。音频和视频都是非常重要的多媒体形式，Android 系统为音频、视频等多媒体的播放、录制提供了强大的支持。通过本章学习，需要重点掌握如何使用 MediaPlayer 类播放音频，如何使用 VideoView 控件播放视频。除此之外，还需要掌握通过 MediaRecorder 类录制音频的方法，以及控制摄像头拍照、录制视频的方法。通过本章知识的学习，希望读者能够开发一些跟音频和视频相关的软件。

习题 5

1. 简述使用 MediaPlayer 播放音频的步骤。
2. 简述 MediaRecorder 录制音频的步骤。
3. 请实现一个简单的音乐播放器，至少包含播放、暂停、上一曲、下一曲按钮以及显示播放时长的进度条。

第6章 数据存储与I/O

本章导图

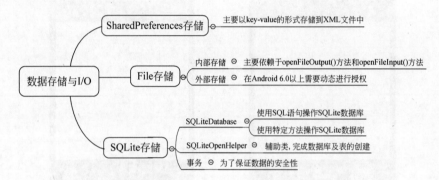

主要内容

- SharedPreferences 存储。
- SharedPreferences 的使用。
- SharedPreferences 的存储位置和格式。
- File 存储。
- 读写 SD 卡上的文件。
- SQLite 数据库。
- 创建数据库和表。
- 使用 SQL 语句操作 SQLite 数据库。
- 使用特定方法操作 SQLite 数据库。
- 事务。
- SQLiteOpenHelper 类。

难点

- SharedPreferences 实现的数据存储功能。
- SharedPreferences 的存储位置和格式。
- SQLite 数据库的使用，实现数据的增、删、改、查。
- 事务。

Android 为开发者提供多种数据存储方式，分别为 SharedPreferences、File 存储、

SQLite 存储、ContentProvider 以及网络存储。开发者在实际开发过程中选择哪种方式依赖于特定需求，需要综合考虑数据存储的类型、需要的空间大小、是否需要提供给其他应用程序使用等多方面因素。本章主要讲解前三种方式，关于 ContentProvider 和网络存储的内容会在后面的章节单独讲解。

6.1 SharedPreferences 存储

SharedPreferences 是 Android 平台上一个轻量级的存储辅助类，它提供了 String、set、int、long、float、boolean 6 种数据类型。当程序中有一些少量数据需要持久化存储时，可以使用 SharedPreferences 类进行存储，如存储程序中的用户名、密码、自定义的一些配置信息等。本节将针对 SharedPreferences 的使用进行详细的讲解。

6.1.1 SharedPreferences 的使用

SharedPreferences 对象本身只能获取数据而不支持存储和修改，存储和修改是通过 SharedPreferences.edit() 方法获取的内部接口 Editor 对象实现的。使用 Preference 来存取数据，用到了 SharedPreferences 接口和 SharedPreferences 的一个内部接口 SharedPreferences.Editor，这两个接口在 android.content 包中。

1. 写数据

使用 SharedPreferences 类写入数据时，首先调用 getSharedPreferences(String name, int mode)方法来创建一个 SharedPreferences 对象，然后调用 SharedPreferences 类的 edit()方法获取可编辑的 Editor 对象，最后通过该对象的 putXxx()方法存储数据。示例代码如下：

```
SharedPreferences sp = getSharedPreferences("info", MODE_PRIVATE);
sharedPreferences.Editor editor = sp.edit();
editor.putstring("name","张三");
editor.putInt("age", 18);
editor.commit();
```

由上述代码可知，getSharedPreferences()方法接收两个参数：第一个参数用于指定 SharedPreferences 文件的名称，如果指定的文件不存在则会创建一个；第二个参数用于指定操作模式，主要有以下几种模式可以选择。

MODE_PRIVATE：默认的操作模式，和直接传入 0 效果是相同的。指定该 SharedPreferences 数据只能被本应用程序读写。

MODE_WORLD_READABLE：指定该 SharedPreferences 数据能被其他应用程序读，但不能写。

MODE_WORLD_WRITEABLE：指定该 SharedPreferences 数据能被其他应用程序读。

MODE_APPEND：该模式会检查文件是否存在，存在就往文件追加内容，否则就创建新文件。

需要注意的是，Editor 对象是以 key-value 的形式保存数据的，并且根据数据类型的不同，会调用不同的方法。操作完数据后，一定要调用 commit()方法进行数据提交，否则所有操作不生效。

2. 读数据

读取 SharedPreferences 中的数据非常简单，只需要获取 SharedPreferences 对象，然后通过该对象的 getXxx()方法根据相应 key 值获取到 value 的值即可。示例代码如下：

```
SharedPreferences sp = getSharedPreferences("info",MODE_PRIVATE);
String data = sp.getString("name","");              // 获取用户名
```

3. 删除数据

如果需要删除 SharedPreferences 中的数据，则只需要调用 Editor 对象的 remove (String key)方法或者 clear()方法即可。示例代码如下：

```
editor.remove("name");              //删除一条数据
editor.clear();                     //删除所有数据
```

6.1.2　SharedPreferences 的存储位置和格式

在使用 SharedPreferences 存储数据时，信息具体存储到哪里呢？我们就通过如图 6.1 所示的交互界面进行讲解。

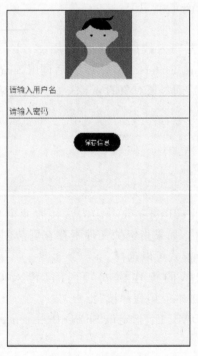

图 6.1　布局界面 1

例 6.1　SaveInfoApp6_1

首先使用一个垂直的线性布局，添加 ImageView、EditText、Button 控件，当输入用户信息时，点击"保存信息"按钮，就会使用 SharedPreferences 进行信息存储。activity_main.xml 的核心代码如下：

```xml
<?xml version="1.0" encoding="utf-8"?>
<LinearLayout xmlns:android="http://schemas.android.com/apk/res/android"
    xmlns:app="http://schemas.android.com/apk/res-auto"
    xmlns:tools="http://schemas.android.com/tools"
    android:layout_width="match_parent"
    android:layout_height="match_parent"
    android:orientation="vertical"
    tools:context=".MainActivity">

    <ImageView
        android:id="@+id/imageView"
        android:layout_width="150dp"
        android:layout_height="150dp"
        android:layout_gravity="center"
        app:srcCompat="@drawable/head"
        tools:srcCompat="@tools:sample/avatars" />

    <EditText
        android:id="@+id/name"
        android:layout_width="match_parent"
        android:layout_height="wrap_content"
        android:ems="10"
        android:hint="请输入用户名"
        android:inputType="text" />

    <EditText
        android:id="@+id/pass"
        android:layout_width="match_parent"
        android:layout_height="wrap_content"
        android:ems="20"
        android:hint="请输入密码"
        android:inputType="textPassword" />

    <Button
        android:id="@+id/btn_save"
        android:layout_width="wrap_content"
        android:layout_height="wrap_content"
        android:layout_marginLeft="150dp"
        android:layout_marginTop="20dp"
        android:text="保存信息" />
</LinearLayout>
```

接下来就是编写交互的 MainActivity,其核心代码如下:

```
package com.example.saveinfoapp6_1;
import androidx.appcompat.app.AppCompatActivity;
import android.content.SharedPreferences;
import android.os.Bundle;
import android.view.View;
import android.widget.Button;
import android.widget.EditText;
import android.widget.Toast;
```

```java
public class MainActivity extends AppCompatActivity {
    private EditText name, pass;
    private Button btnSave;

    @Override
    protected void onCreate(Bundle savedInstanceState) {
        super.onCreate(savedInstanceState);
        setContentView(R.layout.activity_main);
        name = findViewById(R.id.name);
        pass = findViewById(R.id.pass);
        btnSave = findViewById(R.id.btn_save);
        btnSave.setOnClickListener(new View.OnClickListener() {
            @Override
            public void onClick(View view) {
                SharedPreferences sp = getSharedPreferences("info", MODE_PRIVATE);
                SharedPreferences.Editor editor = sp.edit();
                editor.putString("name", name.getText().toString());
                editor.putString("pass", pass.getText().toString());
                editor.commit();
            }
        });
    }
}
```

图 6.2 SaveInfoApp6_1 运行结果

上述代码中,对定义的按钮控件设置监听事件,当点击按钮时,创建一个 SharedPreferences 对象,使用 Editor 对象将信息存储起来,最后不要忘记将事务进行提交。

SaveInfoApp6_1 运行结果如图 6.2 所示。

保存到 SharedPreferences 中的数据,最终将以 XML 文件的形式保存。运行上述程序后,打开 Android Device Monitor,在 File Explorer 窗口下,找到程序对应的文件夹,在 shared_prefs 文件夹下可以发现名为 info.xml 的文件,如图 6.3 所示,该文件以 key-value 的形式保存了存储在 SharedPreferences 中的数据。

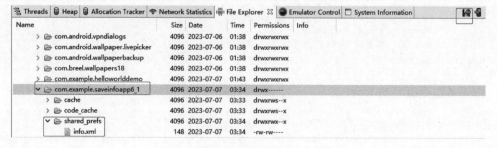

图 6.3 info.xml

6.2 File 存储

File(文件)存储是 Android 系统中一种基本的数据存储方式,与 Java 中的文件存储类似，Android 支持以 I/O 流的形式对文件进行读取和写入操作。在 Android 中有两个文件存储区域：内部存储区和外部存储区。这两种区域的划分源自早期 Android 系统中内置的不可变的内存（Internal Storage）和可卸载的存储部件（External Storage，类似 SD 卡）。虽然现在一些 Android 设备将内部存储区与外部存储区都做成了不可卸载的内置存储，但是这一整块还是从逻辑上被划分为内部存储区与外部存储区，只是现在不再以是否可卸载进行区分了。

1. 内部存储

内部存储是指将应用程序中的数据以文件的形式存储到应用中（该文件默认位于 data/data/< packagename >/目录下），此时存储的文件会被其所在的应用程序私有化，如果其他应用程序想要操作本应用程序中的文件，则需要设置权限。当创建的应用程序被卸载时，其内部存储文件也随之被删除。

Android 开发中，内部存储使用的是 Context 提供的 openFileOutput（）方法和 openFileInput（）方法，这两个方法能够返回进行读写操作的 FileOutputStream 对象和 FileInputStream 对象。如下示例代码将"HelloWorld!"字符串保存到当前应用的内部存储目录下名为 myfile 的文件中：

```
String filename = "myfile.txt";
String data = "HelloWorld!";
FileOutputStream fos = null;
try{
    fos = openFileOutput(filename,Context.MODE_PRIVATE) ;
    fos.write(data.getBytes());
    fos.close();
} catch(Exception e){
    e.printStackTrace();
}
```

上述代码中，首先定义了两个 String 类型的变量 fileName 和 data，这两个变量的值"myfile.txt"与" HelloWorld!"分别表示文件名与要写入文件的数据，接着创建了 FileOutputStream 对象 fos，通过该对象的 write（）方法将数据"HelloWorld!"写入 myfile.txt 文件。

其中，openFileOutput（）方法用于打开应用程序中对应的输出流，将数据存储到指定的文件中。它的参数 name 表示文件名；mode 表示文件的操作模式，也就是读写文件的方式。mode 的取值有 4 种，具体如下：

（1）MODE_PRIVATE：该文件只能被当前程序读写。

（2）MODE_APPEND：该文件的内容可以追加。

（3）MODE_WORLD_READABLE：该文件的内容可以被其他程序读。

（4）MODE_WORLD_WRITEABLE：该文件的内容可以被其他程序写。

需要注意的是，Android 系统有一套自己的安全模型，默认情况下任何应用创建的文件都是私有的，其他程序无法访问，除非在文件创建时指定了操作模式为 MODE_WORLD_READABLE 或者 MODE_WORLD_WRITEABLE。如果希望文件能够被其他程序进行读写操作，则需要同时指定该文件的 MODE_WORLD_READABLE 和 MODE_WORLD_

WRITEABLE 权限。

2. 外部存储

内部存储中的数据对应用来说是私密的,用户和其他应用都没有访问权限,而外部存储中的数据是可以被其他应用或用户访问甚至删除的,用户可以通过 USB 方式和 PC 之间交互外部存储中的数据。外部存储的文件通常位于 mnt/sdcard 目录下,不同厂商生产的手机路径可能会不同。外部存储的文件可以被其他应用程序所共享,当将外部存储设备连接到计算机时,这些文件可以被浏览、修改和删除,因此这种方式不安全。

由于外部存储设备可能被移除、丢失或者处于其他状态,因此在使用外部设备之前必须使用 Environment.getExternalStorageState()方法确认外部设备是否可用,当外部设备可用并且具有读写权限时,那么就可以通过 FileInputStream、FileOutputStream 对象来读写外部设备中的文件。

向外部设备(SD 卡)中存储数据的示例代码如下:

```
String state = Environment.getExternalStorageState();      //获取外部设备的状态
if (state.equals(Environment.MEDIA_MOUNTED)) {             //判断外部设备是否可用
    File SDPath = Environment.getExternalStorageDirectory();  //获取 SD 卡目录
    File file = new File(SDPath,"myfile.txt");
    String data = "HelloWorld!";
    FileOutputStream fos = null;
    try {
        fos = new FileOutputStream(file);
        fos.write(data.getBytes());
    } catch (Exception e) {
        e.printStackTrace();
    } finally {
        try {
            if (fos!= null){
                fos.close();
            }
        }catch(IOException e){
            e.printStackTrace();
        }
    }
}
```

上述代码中,由于手机厂商不同,SD 卡的根目录也不同,为了避免把路径写成固定的值,找不到 SD 卡,需要通过 getExternalStorageDirectory()方法解决此问题。

6.2.1　打开应用中数据文件的 I/O 流

上面讲解了如何将数据以文件的形式写入内部存储和外部存储的文件中。存储好数据之后,如果需要获取这些数据,则需要从文件中读取存储的数据。以下面的例子进行讲解,其布局如图 6.4 所示。

例 6.2　ReadAndWriteApp6_2

首先使用一个相对布局,包含 1 个 EditText 控件用于输入要保存的信息、1 个 TextView 控件用来显示保存的信息和 2 个 Button 控件对应"保存"和"显示"监听事件。其 activity_main.xml 的核心代码如下:

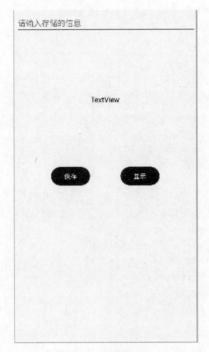

图 6.4 布局界面 2

```xml
<?xml version = "1.0" encoding = "utf-8"?>
<RelativeLayout xmlns:android = "http://schemas.android.com/apk/res/android"
    xmlns:app = "http://schemas.android.com/apk/res-auto"
    xmlns:tools = "http://schemas.android.com/tools"
    android:layout_width = "match_parent"
    android:layout_height = "match_parent"
    tools:context = ".MainActivity">

    <EditText
        android:id = "@+id/input_msg"
        android:layout_width = "match_parent"
        android:layout_height = "wrap_content"
        android:layout_alignParentStart = "true"
        android:layout_alignParentTop = "true"
        android:layout_marginStart = "5dp"
        android:layout_marginTop = "4dp"
        android:ems = "10"
        android:hint = "请输入存储的信息"
        android:inputType = "text" />

    <TextView
        android:id = "@+id/show_msg"
        android:layout_width = "wrap_content"
        android:layout_height = "wrap_content"
        android:layout_alignParentTop = "true"
        android:layout_centerHorizontal = "true"
        android:layout_marginTop = "186dp"
        android:text = "TextView"
```

```xml
            android:textSize = "16sp" />

    < Button
        android:id = "@ + id/btn_save"
        android:layout_width = "wrap_content"
        android:layout_height = "wrap_content"
        android:layout_alignParentStart = "true"
        android:layout_centerVertical = "true"
        android:layout_marginStart = "83dp"
        android:text = "保存" />

    < Button
        android:id = "@ + id/btn_show"
        android:layout_width = "wrap_content"
        android:layout_height = "wrap_content"
        android:layout_alignParentEnd = "true"
        android:layout_centerVertical = "true"
        android:layout_marginEnd = "83dp"
        android:text = "显示" />
</RelativeLayout>
```

接下来就是编写交互的 MainActivity,其核心代码如下:

```java
package com.example.readandwriteapp6_2;
import androidx.appcompat.app.AppCompatActivity;
import android.content.Context;
import android.os.Bundle;
import android.view.View;
import android.widget.Button;
import android.widget.EditText;
import android.widget.TextView;
import java.io.BufferedReader;
import java.io.FileInputStream;
import java.io.FileNotFoundException;
import java.io.FileOutputStream;
import java.io.IOException;
import java.io.InputStreamReader;
import java.nio.charset.StandardCharsets;

public class MainActivity extends AppCompatActivity {
    private EditText inputMsg;
    private Button btnSave, btnShow;
    private TextView showMsg;
    private String filename = "data.txt"; //文件的名称

    @Override
    protected void onCreate(Bundle savedInstanceState) {
        super.onCreate(savedInstanceState);
        setContentView(R.layout.activity_main);
```

```java
        //控件变量的初始化
        inputMsg = findViewById(R.id.input_msg);
        btnSave = findViewById(R.id.btn_save);
        btnShow = findViewById(R.id.btn_show);
        showMsg = findViewById(R.id.show_msg);
        //保存信息的点击事件
        btnSave.setOnClickListener(new View.OnClickListener() {
            @Override
            public void onClick(View view) {
             saveMsg(inputMsg.getText().toString());
            }
        });
        //显示信息的点击事件
        btnShow.setOnClickListener(new View.OnClickListener() {
            @Override
            public void onClick(View view) {
                String msg = readMsg();
                showMsg.setText(msg);
            }
        });
    }

    //保存信息的方法
    private void saveMsg(String fileContents) {
        try (FileOutputStream fos = openFileOutput(filename, Context.MODE_PRIVATE)) {
            fos.write(fileContents.getBytes());
        } catch (FileNotFoundException e) {
            e.printStackTrace();
        } catch (IOException e) {
            e.printStackTrace();
        }
    }
    //读取信息的方法
    private String readMsg() {
        String contents = "";
        try (FileInputStream fis = openFileInput(filename)) {
            InputStreamReader inputStreamReader =
new InputStreamReader(fis, StandardCharsets.UTF_8);
            StringBuilder stringBuilder = new StringBuilder();
            try (BufferedReader reader = new BufferedReader(inputStreamReader)) {
                String line = reader.readLine();
                while (line != null) {
                    stringBuilder.append(line).append('\n');
                    line = reader.readLine();
                }
            } catch (IOException e) {

            } finally {
                contents = stringBuilder.toString();
            }
        } catch (IOException e) {
            e.printStackTrace();
        }
        return contents;
    }
}
```

上述代码中,实现了两个方法:saveMsg()用于将从输入框 input_msg 中获取的信息保存到内部文件 data.txt 中,当点击界面上的"保存"按钮时,调用该方法;readMsg()用于从内部文件 data.txt 中读取信息,当点击界面上的"显示"按钮时,调用该方法,最后将获取的信息显示到控件 showMsg 上。对应文件的读取操作在此不再赘述。其运行结果如图 6.5 所示。

图 6.5 ReadAndWriteApp6_2 运行结果

6.2.2 读写 SD 卡上的文件

在例 6.2 中,讲解了如何读取内部存储的文件信息,然而 SD 卡作为手机的扩展存储设备,在手机中充当硬盘角色,可以让手机存放更多的数据以及多媒体等大体积文件。因此查看 SD 卡的内存就跟查看硬盘的剩余空间一样,那么在 Android 开发中,如何能获取 SD 卡的文件信息呢?接下来还是以一个例子进行讲解。

例 6.3 ReadSDApp6_3

为了简化代码量,我们在此采用例 6.2 所使用的布局文件,因此代码在此就不再展示。

为了保证应用程序的安全性,Android 系统规定,程序访问系统的一些关键信息时,必须申请权限。否则程序运行时会因为没有访问系统信息的权限而直接崩溃。根据程序适配的 Android SDK 版本的不同,申请权限分为两种方式,分别为静态申请权限和动态申请权限。静态申请权限的方式适用于 Android SDK 6.0 以下的版本。该方式是在清单文件(AndroidManifest.xml)的< manifest >节点中声明需要申请的权限。示例代码如下:

```
< uses - permission android:name = "android.permission.WRITE_EXTERNAL_STORAGE" />
```

动态申请权限:当程序适配的 Android SDK 版本为 6.0 及以上时,Android 改变了权限的管理模式,权限被分为正常权限和危险权限。具体如下：

正常权限：表示不会直接给用户隐私权带来风险的权限,如请求网络的权限。

危险权限：表示涉及用户隐私的权限,申请了该权限的应用,可能涉及了用户隐私信息的数据或资源,也可能对用户存储的数据或其他应用的操作产生影响。危险权限一共有9组,分别为位置(LOCATION)、日历(CALENDAR)、照相机(CAMERA)、联系人(CONTACTS)、存储卡(STORAGECARD)、传感器(SENSORS)、麦克风(MICROPHONE)、电话(PHONE)和短信(SMS)的相关权限。

申请正常权限时使用静态申请权限的方式即可,但是对于一些涉及用户隐私的危险权限需要用户的授权后才可以使用,因此危险权限不仅需要在清单文件(AndroidManifest.xml)的< manifest >节点中添加权限,在该App的manifests的AndroidManifest.xml文件中添加读写SD卡的权限：

```
< uses-permission android:name = "android.permission.WRITE_EXTERNAL_STORAGE" />
< uses-permission android:name = "android.permission.READ_EXTERNAL_STORAGE" />
```

在Android应用程序获得读写存储卡权限的情况下,还必须在AndroidManifest.xml的application标签下声明requestLegacyExternalStorage＝true,才可以访问沙盒路径下的数据。另外,还需要在代码中动态申请权限。

PermissionListener.java：

```java
package com.example.readsdapp6_3;
import java.util.List;
public interface PermissionListener {
    //授权成功
    void onGranted();
    //授权部分
    void onGranted(List< String > grantedPermission);
    //拒绝授权
    void onDenied(List< String > deniedPermission);
}
```

PermissionActivity.java：

```java
package com.example.readsdapp6_3;
import android.content.pm.PackageManager;
import androidx.appcompat.app.AppCompatActivity;
import androidx.annotation.NonNull;
import androidx.core.content.ContextCompat;
import java.util.ArrayList;
import java.util.List;
public class PermissionActivity extends AppCompatActivity {
    private PermissionListener mlistener;

    /**
     * 权限申请
     * @param permissions 待申请的权限集合
     * @param listener 申请结果监听事件
```

```java
         */
        protected void requestRunTimePermission(String[] permissions, PermissionListener listener){
            this.mlistener = listener;

            //用于存放未授权的权限
            List<String> permissionList = new ArrayList<>();
            //遍历传递过来的权限集合
            for (String permission : permissions) {
                //判断是否已经授权
                if (ContextCompat.checkSelfPermission(this,permission) != PackageManager.PERMISSION_GRANTED){
                    //未授权,则加入待授权的权限集合中
                    permissionList.add(permission);
                }
            }

            //判断集合
            if (!permissionList.isEmpty()){ //如果集合不为空,则需要去授权
                this.requestPermissions(permissionList.toArray(new String[permissionList.size()]),1);

            }else{ //若为空,则已经全部授权
                listener.onGranted();
            }
        }

        /**
         * 权限申请结果
         * @param requestCode       请求码
         * @param permissions       所有的权限集合
         * @param grantResults      授权结果集合
         */
        @Override
        public void onRequestPermissionsResult(int requestCode, @NonNull String[] permissions, @NonNull int[] grantResults) {
            super.onRequestPermissionsResult(requestCode, permissions, grantResults);
            switch (requestCode) {
                case 1:
                    if (grantResults.length > 0){
                        //被用户拒绝的权限集合
                        List<String> deniedPermissions = new ArrayList<>();
                        //用户通过的权限集合
                        List<String> grantedPermissions = new ArrayList<>();
                        for (int i = 0; i < grantResults.length; i++) {
                            //获取授权结果,这是一个int类型的值
                            int grantResult = grantResults[i];

                            if (grantResult != PackageManager.PERMISSION_GRANTED){
                                //用户拒绝授权的权限
                                String permission = permissions[i];
                                deniedPermissions.add(permission);
                            }else{ //用户同意的权限
```

```
                        String permission = permissions[i];
                        grantedPermissions.add(permission);
                    }
                }
                if (deniedPermissions.isEmpty()){  //用户拒绝权限为空
                    mlistener.onGranted();
                }else {  //不为空
                    //回调授权成功的接口
                    mlistener.onDenied(deniedPermissions);
                    //回调授权失败的接口
                    mlistener.onGranted(grantedPermissions);
                }
            }
            break;
        default:
            break;
    }
}
```

在上述代码中，PermissionListener.java 接口中定义了三个抽象方法，主要是为了监视授权是否成功。requestPermissions()方法中包含 2 个参数：第 1 个参数为需要申请的权限；第 2 个参数为接口回调。添加完动态申请权限后，运行程序，界面上会弹出是否允许请求权限的对话框，由用户进行授权，如图 6.6 所示。

图 6.6　申请授权对话框

在图 6.6 中，提示内容为"是否允许访问设备上照片、媒体和文件的申请权限"，DENY 表示拒绝、ALLOW 表示允许。当用户点击对话框中的 ALLOW 按钮时，程序会执行动态申请权限的回调方法 onRequestPermissionsResult()，在该方法中包含 3 个参数，requestCode、permissions 和 grantResults 分别表示请求码、请求的权限和用户授予权限的结果。当用户授予 SD 卡写权限时，对应该权限的 grantResults 数组中的值为 PackageManager.PERMISSIONGRANTED。

MainActivity.java：

```
package com.example.readsdapp6_3;
import android.Manifest;
import android.os.Bundle;
import android.os.Environment;
import android.view.View;
import android.widget.Button;
import android.widget.EditText;
import android.widget.TextView;
import java.io.BufferedReader;
import java.io.File;
import java.io.FileInputStream;
import java.io.FileOutputStream;
import java.io.IOException;
```

```java
import java.io.InputStreamReader;
import java.nio.charset.StandardCharsets;
import java.util.List;

public class MainActivity extends PermissionActivity {
    private EditText inputMsg;
    private Button btnSave, btnShow;
    private TextView showMsg;
    private boolean isGranted = true;              //监视授权是否成功

    @Override
    protected void onCreate(Bundle savedInstanceState) {
        super.onCreate(savedInstanceState);
        setContentView(R.layout.activity_main);
        //控件变量的初始化
        inputMsg = findViewById(R.id.input_msg);
        btnSave = findViewById(R.id.btn_save);
        btnShow = findViewById(R.id.btn_show);
        showMsg = findViewById(R.id.show_msg);
        //保存信息的点击事件
        btnSave.setOnClickListener(new View.OnClickListener() {
            @Override
            public void onClick(View view) {
                saveMsg(inputMsg.getText().toString());
            }
        });
        //显示信息的点击事件
        btnShow.setOnClickListener(new View.OnClickListener() {
            @Override
            public void onClick(View view) {
                String msg = readMsg();
                showMsg.setText(msg);
            }
        });

    }
    //保存信息的方法
    private void saveMsg(String fileContents) {
        requestPermission();
        if (isGranted) {
            File sdcard = Environment.getExternalStorageDirectory();//获取 SD 卡的路径
            File file = new File(sdcard, "myfile.txt");                //创建 myfile.txt 文件
            try (FileOutputStream fos = new FileOutputStream(file)) {
                fos.write(fileContents.getBytes());
            } catch (IOException e) {
                e.printStackTrace();
            }
        }
    }
    //读取信息的方法
    private String readMsg() {
        requestPermission();
        String contents = "";
```

```java
            if (isGranted) {
                File sdcard = Environment.getExternalStorageDirectory();
                File file = new File(sdcard, "myfile.txt");
                StringBuilder stringBuilder = new StringBuilder();
                try (FileInputStream fis = new FileInputStream(file)) {
                    InputStreamReader inputStreamReader = new InputStreamReader(fis,
StandardCharsets.UTF_8);
                    BufferedReader reader = new BufferedReader(inputStreamReader);
                    String line = reader.readLine();
                    while (line != null) {
                        stringBuilder.append(line).append('\n');
                        line = reader.readLine();
                    }

                } catch (IOException e) {
                    e.printStackTrace();
                } finally {
                    contents = stringBuilder.toString();
                }
            }
            return contents;
        }
        //请求授权的方法
        private void requestPermission() {
            String[] permissionList = new String[]{Manifest.permission.WRITE_EXTERNAL_STORAGE,
                    Manifest.permission.READ_EXTERNAL_STORAGE};
            requestRunTimePermission(permissionList, new PermissionListener() {
                @Override
                public void onGranted() {
                    isGranted = true;
                }
                @Override
                public void onGranted(List<String> grantedPermission) {

                }
                @Override
                public void onDenied(List<String> deniedPermission) {
                    isGranted = false;
                }
            });
        }
    }
```

在上述代码中，MainActivity 类继承 PermissionActivity 类，首先定义了一个方法 requestPermission()用来请求授权，如果授权成功则全局变量 isGranted 值设置为 true；其次定义了两个方法 saveMsg()和 readMsg()，用来读写 SD 上的信息，在执行这两个方法前，需要调用请求授权的方法，先判断授权是否成功，如果授权成功则继续执行代码。ReadSDApp6_3 运行结果如图 6.7 所示。

图 6.7　ReadSDApp6_3 运行结果

6.3　SQLite 存储

前面介绍了如何使用 File 和 SharedPreferences 存储数据,这两种方式只适合存储少量简单数据,如果需要存储大量数据,显然不合适。为此,Android 系统提供了 SQLite 数据库,它是一个轻量级数据库,占用资源非常低,在内存中只需要占用几十万字节的存储空间。SQLite 是遵守 ACID 关联式的数据库管理系统,没有服务器进程,它通过文件保存数据,该文件是跨平台的,可以放在其他平台中使用,并且支持 NULL、INTEGER、REAL(浮点数字)、TEXT(字符串文本)和 BLOB(二进制对象)5 种数据类型。

6.3.1　SQLiteDatabase 简介

Android 提供了一个名为 SQLiteDatabase 的类,该类封装了一些操作数据库的 API,使用该类可以完成数据的添加(Create)、查询(Retrieve)、更新(Update)和删除(Delete)操作(这些操作简称为 CRUD)。它等同于 JDBC 中 Connection 和 Statement 的结合体。SQLiteDatabase 既代表与数据库的连接,又能用于执行 SQL 语句操作。其常用的方法如表 6.1 所示。

表 6.1　SQLiteDatabase 常用的方法

方　　法	说　　明
create(SQLiteDatabase.CursorFactory factory)	创建数据库
openOrCreateDatabase(File file, SQLiteDatabase.CursorFactory factory)	创建或打开数据库

续表

方　　法	说　　明
openOrCreateDatabase（String path，SQLiteDatabase.CursorFactory factory）	创建或打开数据库
public long insert（String table，String nullColumnHack，ContentValues values）	向指定的数据表中添加一条记录
public Cursor query（String table，String［］columns，String selection，String selectionArgs，String groupBy，String having，String orderBy）	用于查询数据
public Cursor rawQuery(String sql，Strings[] selectionArgs)	执行带占位符的 SQL 查询
public int update（String table，ContentValues values，String whereClause，String[] whereArgs）	修改特定数据
public int delete（String table，String whereClause，String whereArgs）	删除表中特定的记录
public votd execSQL(String sql，Object[] bindArgs)	执行一条带有占位的 SQL 语句
public void close()	关闭数据库

需要注意的是，SQLiteDatabase 的构造方法是 private()，因此只能通过 openDatabase()来连接数据库；参数中的 path 代表数据库的路径（如果是在默认路径/data/data/<package_name>/databases/下，则这里只需要提供数据库名称）；factory 代表在创建 Cursor 对象时使用的工厂类，如果为 null，则使用默认的工厂。

6.3.2　SQLiteOpenHelper 类

SQLiteOpenHelper 类是 SQLiteDatabase 的一个辅助类。这个类的主要作用是生成一个数据库，并对数据库的版本进行管理。在实际开发过程中较少直接使用 SQLiteDatabase 的方法打开数据库，通常会继承 SQLiteOpenHelper 开发子类，并通过该类的 getReadableDatabase()方法或 getWritableDatabase()方法打开数据库。

SQLiteOpenHelper 类常用的方法如表 6.2 所示。

表 6.2　**SQLiteOpenHelper 类常用的方法**

方　　法	说　　明
public SQLiteOpenHelper（Context context，String name，CursorFactory factory，int version）	构造方法
public void onCreate(SQLiteDatabase db)	在数据库第一次生成时会调用这个方法
public void onUpgrade（SQLiteDatabase db，int oldVersion，int newVersion）	数据库版本更新时调用
public SQLiteDatabase getReadableDatabase()	创建或打开一个只读数据库
public SQLiteDatabase getWritableDatabase()	创建或打开一个读写的数据库

SQLiteOpenHelper 类根据开发应用程序的需要，封装了创建和更新数据库使用的逻辑。定义类继承 SQLiteOpenlHelper，至少需要在类中实现如下 3 个方法。

（1）构造方法。需要在构造方法中调用父类 SQLiteOpenHelper 的构造方法。

（2）onCreate（SQLiteDatabase db）方法。应用第一次使用时会调用 onCreate

(SQLiteDatabase db)方法生成数据库中的表,即在程序中调用方法 getWriteableDatabase()或 getReadableDatabase()方法获取用于操作数据库的 SQLiteDatabase 实例时,如果数据库不存在,Android 系统会自动生成一个数据库,然后调用 onCrcate()方法用于生成数据库的表。因为 onCreate()方法在初次生成数据库时才调用,重写 onCreate()方法时,可以生成数据表结构以及添加数据到数据库中。

(3) onUpgrade(SQLiteDatabase db,int oldVersion,int newVersion)方法。用于升级软件时更新数据表结构,该方法在数据库的版本发生变化时调用。该方法中参数 oldVersion 代表数据库之前的版本号,参数 newVersion 代表数据库当前的版本号。

6.3.3 创建数据库和表

通过上面知识的介绍可知,在 Android 中,通常会通过继承 SQLiteOpenHelper 类并重写 onCreate()方法来创建数据库和表。创建数据库和表的示例代码如下所示:

```java
public class ContactDBHelper extends SQLiteOpenHelper {
    public ContactDBHelper(@Nullable Context context) {
        super(context, "information.db", null, 1);
    }

    @Override
    public void onCreate(SQLiteDatabase db) {
        db.execSQL("CREATE TABLE " +
                "user(_id INTEGER PRIMARY KEY AUTOINCREMENT ," +
                "name VACHAR(20),phone VARCHAR(20))");
    }
    @Override
    public void onUpgrade(SQLiteDatabase sqLiteDatabase, int i, int i1) {

    }
}
```

在上述示例代码中,定义了 SQLiteOpenHelper 的子类 ContactDBHelper,在子类构造方法中通过 super()方法调用了父类的构造方法创建了名为 information.db 的数据库,其中参数代表上下文、数据库的名称、工厂模式(一般为 null)、数据库的版本。使用 onCreate()方法完成了表 user 的创建,只有当数据库的版本发生更新时,才会执行 onUpgrade()方法。

6.3.4 使用 SQL 语句操作 SQLite 数据库

很多读者在前期的课程中学习过 SQL,会写 SQL 语句,而不想使用 Android 提供的 API,那么就可以利用 Android 提供的相关方法去执行这些 SQL 语句。例如定义一个 Person 类,有 id、name、phone 这三个属性,SQLite 数据库中有一张 person 表,下面根据这些条件讲解使用 SQL 语句操作 SQLite 数据库。

1. 插入数据

```java
public void save(Person p)
{
    SQLiteDatabase db = dbOpenHelper.getWritableDatabase();
```

```java
    String sql = "INSERT INTO person(name,phone) values(?,?)";
    Object []obj={p.getName(),p.getPhone()};
    db.execSQL(sql,obj);
    db.close();
}
```

2. 删除数据

```java
public void delete(Integer id)
{
    SQLiteDatabase db = dbOpenHelper.getWritableDatabase();
    String sql = "DELETE FROM person WHERE personid = ?";
    Object[] obj = {id};
    db.execSQL(sql,obj);
    db.close();
}
```

3. 修改数据

```java
public void update(Person p)
{
    SQLiteDatabase db = dbOpenHelper.getWritableDatabase();
    String sql = "UPDATE person SET name = ?,phone = ? WHERE personid = ?";
    Object[] obj = {p.getName(),p.getPhone(),p.getId()};
    db.execSQL(sql,obj);
    db.close();
}
```

4. 根据 id 查询单条数据

```java
public Person find(Integer id)
{
    SQLiteDatabase db = dbOpenHelper.getReadableDatabase();
    String sql = "SELECT * FROM person WHERE personid = ?";
    String[] obj = {id.toString()};
    Cursor cursor = db.rawQuery(sql,obj);
    //存在数据才返回 true
    if(cursor.moveToFirst())
    {
        int personid = cursor.getInt(cursor.getColumnIndex("id"));
        String name = cursor.getString(cursor.getColumnIndex("name"));
        String phone = cursor.getString(cursor.getColumnIndex("phone"));
        return new Person(id,name,phone);
    }
    cursor.close();
    return null;
}
```

5. 查询记录数

```java
public long getCount()
{
```

```
    SQLiteDatabase db = dbOpenHelper.getReadableDatabase();
    String sql = "SELECT COUNT ( * ) FROM person";
    Cursor cursor = db.rawQuery(sql,null);
    cursor.moveToFirst();
    long result = cursor.getLong(0);
    cursor.close();
    return result;
}
```

6.3.5 使用特定方法操作 SQLite 数据库

对于 SQLite 数据库的操作,除了使用 SQL 语句之外,Android 还提供了一些特定的方法来进行操作,这些方法更加方便和安全。还以 6.3.4 节给定的前提条件,讲解如何使用特定的方法实现数据的增加、删除、修改、查询操作。

1. 插入数据

```
public void insert(Person p)
{
    SQLiteDatabase db = dbOpenHelper.getWritableDatabase();
    ContentValues values = new ContentValues();
    values.put(p.getName(), p.getPhone());
    long newRowId = db.insert("person",null,values);
    db.close();
}
long insert(String table,String nullColumnHack,ContentValues values)
```

insert:方法签名。

table:表名。

nullColumnHack:强行插入 null 值的数据列的列名。当 values 为 null 或它包含的键值对的数量为 0 时,就起作用了。它不应是主键列的列名,也不应是非空列的列名。

values:代表一行记录的数据。ContentValues 类似于 Map 集合,存放键值对,键为数据列的列名。

2. 删除数据

```
public void delete(Integer id)
{
    SQLiteDatabase db = dbOpenHelper.getWritableDatabase();
    int deletedRows = db.delete("person","id = ?",new String[]{id.toString()});
    db.close();
}
int delete(String table, String whereClause,String[] whereArgs)
```

delete:方法签名。

table:表名。

whereClause:条件,满足此条件的记录将会被删除。

whereArgs:用于为 whereClause 子句传入参数,即替代占位符。

返回的整数是受此 delete 语句影响的记录的条数。

3. 修改数据

```
public void update(Person p)
{
    SQLiteDatabase db = dbOpenHelper.getWritableDatabase();
    ContentValues values = new ContentValues();
    values.put(p.getName(), p.getPhone());
    int updateRows = db.update("person",values,"id = ?",new String[]{p.getId()});
    db.close();
}
int update(String table,ContentValues values,String whereClause,String[] whereArgs)
```

update：方法签名。

table：表名。

values：想更新的数据。

whereClause：条件。

whereArgs：为 whereClause 子句传入参数，即用来替代占位符的内容。

返回的整数是受此 update 语句影响的记录的条数。

4. 查询数据

```
public void query()
{
    SQLiteDatabase db = dbOpenHelper.getReadableDatabase();
    Cursor cursor = db.query("person",null, null, null,
            null, null,null);
    List<Person> personList = new ArrayList<Person>();
    while(cursor.moveToNext()) {
        int id = cursor.getInt(0);
        String name = cursor.getString(1);
        String phone = cursor.getString(2);
        Person p = new Person(id,name,phone);
        personList.add(p);
    }
    cursor.close();
    db.close();
}
cursor query(String table,String[] columns,String selection,String[] selectionArgs,String groupBy, String orderBy)
```

query：方法签名。

table：表名。

columns：要查询的列名。

selection：查询条件子句，相当于 where 关键字后面的部分。

selectionArgs：为 selection 传入参数，替代占位符。

groupBy：控制分组，相当于 select 语句 group by 后面的部分。

having：用于对分组进行过滤，相当于 select 语句 having 后面的部分。

orderBy：排序，相当于 select 语句 order by 后面的部分。

6.3.6 事务

事务(Transaction)是一个对数据库执行工作单元。事务是以逻辑顺序完成的工作单位或序列,可以由用户手动操作完成,也可以由某种数据库程序自动完成。

事务是指一个或多个更改数据库的扩展。例如,如果您正在创建一个记录或者更新一个记录或者从表中删除一个记录,那么您正在该表上执行事务。重要的是,要控制事务以确保数据的完整性和处理数据库错误。

实际上,可以把许多的 SQLite 查询联合成一组,把所有这些放在一起作为事务的一部分进行执行。

1. 事务的属性

事务具有以下四个标准属性,通常根据首字母缩写为 ACID。

原子性(Atomicity):确保工作单位内的所有操作都成功完成,否则,事务会在出现故障时终止,之前的操作也会回滚到以前的状态。

一致性(Consistency):确保数据库在成功提交的事务上正确地改变状态。

隔离性(Isolation):使事务操作相互独立和透明。

持久性(Durability):确保已提交事务的结果或效果在系统发生故障的情况下仍然存在。

2. 事务控制

事务控制命令只与 DML 命令 insert、update 和 delete 一起使用。它们不能在创建表或删除表时使用,因为这些操作在数据库中是自动提交的。

begin transaction 命令:事务可以使用 begin transaction 命令或简单的 begin 命令来启动。此类事务通常会持续执行下去,直到遇到下一个 commit 或 rollback 命令。不过在数据库关闭或发生错误时,事务处理也会回滚。以下是启动一个事务的简单语法:

```
begin;
```

或

```
begin transaction;
```

commit 命令:用于把事务调用的更改保存到数据库中的事务命令。commit 命令把自上次 commit 或 rollback 命令以来的所有事务保存到数据库。commit 命令的语法如下:

```
commit;
```

或

```
end transaction;
```

rollback 命令:用于撤销尚未保存到数据库的事务命令。rollback 命令只能用于撤销自上次发出 commit 或 rollback 命令以来的事务。rollback 命令的语法如下:

```
rollback;
```

6.4 应用实例：手机通讯录

在上面讲解了 SQLite 数据库的创建以及基本操作，接下来通过一个手机通讯录的实例对 SQLite 数据库在开发中的应用进行详细讲解。该程序能够对联系人和电话进行添加、删除、修改和查询操作，并且通讯录信息以列表的形式展示。交互界面如图 6.8 所示，其具体实现的核心代码如下。

观看视频

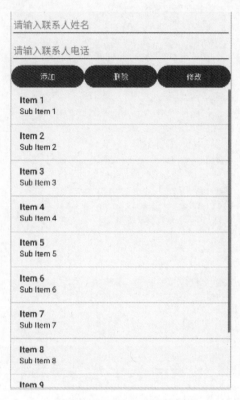

图 6.8 交互界面

例 6.4 ContactApp6_4

根据图 6.8 所示的交互界面，编写的 activity_main.xml 布局文件的代码如下：

```xml
<?xml version = "1.0" encoding = "utf-8"?>
<RelativeLayout xmlns:android = "http://schemas.android.com/apk/res/android"
    xmlns:app = "http://schemas.android.com/apk/res-auto"
    xmlns:tools = "http://schemas.android.com/tools"
    android:layout_width = "match_parent"
    android:layout_height = "match_parent"
    tools:context = ".MainActivity">

    <EditText
        android:id = "@ + id/input_name"
        android:layout_width = "match_parent"
        android:layout_height = "wrap_content"
```

```xml
        android:layout_alignParentStart = "true"
        android:layout_alignParentTop = "true"
        android:layout_marginStart = "0dp"
        android:ems = "10"
        android:hint = "请输入联系人姓名"
        android:inputType = "text" />

    <EditText
        android:id = "@+id/input_phone"
        android:layout_width = "match_parent"
        android:layout_height = "wrap_content"
        android:layout_below = "@+id/input_name"
        android:layout_alignParentStart = "true"
        android:ems = "15"
        android:hint = "请输入联系人电话"
        android:inputType = "text" />

    <LinearLayout
        android:id = "@+id/ly_linear"
        android:layout_width = "match_parent"
        android:layout_height = "wrap_content"
        android:layout_below = "@+id/input_phone"
        android:layout_alignParentStart = "true"
        android:orientation = "horizontal">

        <Button
            android:id = "@+id/btn_insert"
            android:layout_width = "wrap_content"
            android:layout_height = "wrap_content"
            android:layout_weight = "1"
            android:text = "添加" />

        <Button
            android:id = "@+id/btn_delete"
            android:layout_width = "wrap_content"
            android:layout_height = "wrap_content"
            android:layout_weight = "1"
            android:text = "删除" />

        <Button
            android:id = "@+id/btn_update"
            android:layout_width = "wrap_content"
            android:layout_height = "wrap_content"
            android:layout_weight = "1"
            android:text = "修改" />
    </LinearLayout>

    <ListView
        android:id = "@+id/list_view"
        android:layout_width = "match_parent"
        android:layout_height = "match_parent"
        android:layout_below = "@+id/ly_linear"
        android:layout_alignParentStart = "true" />
</RelativeLayout>
```

页面中包含了 2 个 EditText 控件用于输入联系人的姓名和电话，3 个 Button 按钮用于对应"添加""删除""修改"事件，ListView 用于显示数据库中存储的联系人信息。另外，还需要在 res/layout 文件夹下创建一个名为 layout_item.xml 的布局文件，用来控制列表项的显示样式，布局如图 6.9 所示。其核心代码如下：

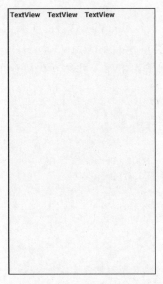

图 6.9　列表项布局

```xml
<?xml version = "1.0" encoding = "utf-8"?>
<LinearLayout xmlns:android = "http://schemas.android.com/apk/res/android"
    android:layout_width = "match_parent"
    android:layout_height = "match_parent"
    android:padding = "5dp">

    <TextView
        android:id = "@+id/text_id"
        android:layout_width = "wrap_content"
        android:layout_height = "wrap_content"
        android:paddingEnd = "20dp"
        android:text = "TextView"
        android:textSize = "20sp"
        android:textStyle = "bold" />

    <TextView
        android:id = "@+id/text_name"
        android:layout_width = "wrap_content"
        android:layout_height = "wrap_content"
        android:text = "TextView"
        android:textSize = "20sp"
        android:paddingEnd = "20dp"
        android:textStyle = "bold" />

    <TextView
        android:id = "@+id/text_phone"
        android:layout_width = "match_parent"
```

```xml
            android:layout_height = "wrap_content"
            android:text = "TextView"
            android:textSize = "20sp"
            android:textStyle = "bold" />
</LinearLayout>
```

列表项布局文件中只包括了 3 个 TextView 控件,主要用来显示编号、姓名、电话。接下来需要创建一个 ContactDBHelper 类继承自 SQLiteOpenHelper 类,在 ContactDBHelper 类中完成数据库、数据表、数据的基本操作工作。其核心代码如下:

```java
package com.example.contactapp6_4;
import android.content.ContentValues;
import android.content.Context;
import android.database.Cursor;
import android.database.sqlite.SQLiteDatabase;
import android.database.sqlite.SQLiteOpenHelper;
import android.widget.Toast;
import androidx.annotation.Nullable;
import java.util.ArrayList;
import java.util.HashMap;
import java.util.List;
import java.util.Map;
import java.util.Objects;

public class ContactDBHelper extends SQLiteOpenHelper {
    private Context context;
    public ContactDBHelper(@Nullable Context context) {
        super(context, "information", null, 1);
        this.context = context;
    }
    @Override
    public void onCreate(SQLiteDatabase db) {
        db.execSQL("CREATE TABLE " +
                "user(_id INTEGER PRIMARY KEY AUTOINCREMENT ," +
                "name VACHAR(20),phone VARCHAR(20))");
    }
    @Override
    public void onUpgrade(SQLiteDatabase sqLiteDatabase, int i, int i1) {

    }
    //添加方法
    public long insert(String name,String phone){
        SQLiteDatabase db = this.getWritableDatabase();
        ContentValues contentValues = new ContentValues();
        contentValues.put("name",name);
        contentValues.put("phone",phone);
        long id = db.insert("user",null,contentValues);
        if (id > 0){
            Toast.makeText(context,"添加成功",Toast.LENGTH_LONG).show();
        }else {
            Toast.makeText(context,"添加失败",Toast.LENGTH_LONG).show();
        }
```

```java
        return id;
    }
    //查询方法
    public List query(){
        SQLiteDatabase db = this.getReadableDatabase();
        Cursor cursor = db.query("user",null,null,null,
                null,null,"name");
        if (cursor.getCount() == 0) {
            cursor.close();
            db.close();
            return null;
        } else {
            List<Map<String,Object>> list = new ArrayList<>();
            while(cursor.moveToNext()){
                Map map = new HashMap();
                map.put("id",cursor.getInt(0));
                map.put("name",cursor.getString(1));
                map.put("phone",cursor.getString(2));
                list.add(map);
            }
            cursor.close();
            db.close();
            return list;
        }
    }
    //删除方法
    public int delete(String name){
        SQLiteDatabase db = getWritableDatabase();
        int id = db.delete("user","name = ?",new String[]{name});
        if (id>0){
            Toast.makeText(context,"添加成功",Toast.LENGTH_LONG).show();
        }else {
            Toast.makeText(context,"添加失败",Toast.LENGTH_LONG).show();
        }
        return id;
    }
    //修改操作
    public int update(String name,String phone){
        SQLiteDatabase db = getWritableDatabase();
        ContentValues contentValues = new ContentValues();
        contentValues.put("name",name);
        contentValues.put("phone",phone);
        int id = db.update("user",contentValues,"name = ?",new String[]{name});
        if (id>0){
            Toast.makeText(context,"修改成功",Toast.LENGTH_LONG).show();
        }else {
            Toast.makeText(context,"修改失败",Toast.LENGTH_LONG).show();
        }
        return id;
    }
}
```

上述代码创建了一个名为 information 的数据库，接着又创建了一张 user 数据表，实现了查询、添加、删除、修改 4 个基本功能。最后编写交互的代码 MainActivity，其核心代码如下：

```java
package com.example.contactapp6_4;
import androidx.appcompat.app.AppCompatActivity;
import android.app.Person;
import android.content.ContentValues;
import android.database.Cursor;
import android.database.sqlite.SQLiteDatabase;
import android.os.Bundle;
import android.view.View;
import android.widget.Button;
import android.widget.EditText;
import android.widget.ListView;
import android.widget.SimpleAdapter;
import java.util.ArrayList;
import java.util.HashMap;
import java.util.List;

public class MainActivity extends AppCompatActivity {
    private EditText inputName,inputPhone;
    private Button btnInsert,btnDelete,btnUpdate;
    private ListView listView;
    private ContactDBHelper contactDBHelper;
    private List<HashMap<String,Object>> list;
    @Override
    protected void onCreate(Bundle savedInstanceState) {
        super.onCreate(savedInstanceState);
        setContentView(R.layout.activity_main);
        //初始化操作
        inputName = findViewById(R.id.input_name);
        inputPhone = findViewById(R.id.input_phone);
        btnInsert = findViewById(R.id.btn_insert);
        btnDelete = findViewById(R.id.btn_delete);
        btnUpdate = findViewById(R.id.btn_update);
        listView = findViewById(R.id.list_view);
        contactDBHelper = new ContactDBHelper(this);
        list = contactDBHelper.query();
        if (list!= null){
            SimpleAdapter adapter = new SimpleAdapter(
                    this,
                    list,
                    R.layout.layout_item,
                    new String[]{"id","name","phone"},
                    new int[]{R.id.text_id,R.id.text_name,R.id.text_phone});
            listView.setAdapter(adapter);
            adapter.notifyDataSetChanged();
        }
        //添加信息按钮
        btnInsert.setOnClickListener(new View.OnClickListener() {
            @Override
            public void onClick(View view) {
```

```java
                    long id = contactDBHelper.insert(inputName.getText().toString(),
inputPhone.getText().toString());
                    if(id>0){
                        list = contactDBHelper.query();
                        SimpleAdapter adapter = new SimpleAdapter(
                                MainActivity.this,
                                list,
                                R.layout.layout_item,
                                new String[]{"id","name","phone"},
                                new int[]{R.id.text_id,R.id.text_name,R.id.text_phone});
                        listView.setAdapter(adapter);
                    }
                    inputName.setText("");
                    inputPhone.setText("");
                }
            });
            //删除信息按钮
            btnDelete.setOnClickListener(new View.OnClickListener() {
                @Override
                public void onClick(View view) {
                    int id = contactDBHelper.delete(inputName.getText().toString());
                    if (id>0){
                        list = contactDBHelper.query();
                        SimpleAdapter adapter = new SimpleAdapter(MainActivity.this,
                                list,
                                R.layout.layout_item,
                                new String[]{"id","name","phone"},
                                new int[]{R.id.text_id,R.id.text_name,R.id.text_phone});
                        listView.setAdapter(adapter);
                    }
                    inputName.setText("");
                }
            });
            //修改信息按钮
            btnUpdate.setOnClickListener(new View.OnClickListener() {
                @Override
                public void onClick(View view) {
                    int id = contactDBHelper.update(inputName.getText().toString(),
inputPhone.getText().toString());
                    if (id>0){
                        list = contactDBHelper.query();
                        SimpleAdapter adapter = new SimpleAdapter(MainActivity.this,
                                list,
                                R.layout.layout_item,
                                new String[]{"id","name","phone"},
                                new int[]{R.id.text_id,R.id.text_name,R.id.text_phone});
                        listView.setAdapter(adapter);
                    }
                    inputName.setText("");
                    inputPhone.setText("");
                }

            });
        }
}
```

该程序运行结果如图 6.10 所示,当输入联系人姓名和电话,点击"添加"按钮时,就会向数据库中添加一条信息并实时显示到 ListView 控件上;输入要删除信息的"姓名",点击"删除"按钮就会删除一条信息,ListView 就会重新刷新信息;同理,更新信息也是如此。

图 6.10 程序运行结果

6.5 小结

- SharedPreferences 存储数据时以 key-value 形式存储到一个 XML 文件中。
- Android 开发中,内部存储使用的是 Context 提供的 openFileOutput()方法和 openFileInput()方法进行访问。
- Android 开发中,外部存储数据的访问需要动态授权。
- SQLite 数据库及表的创建主要是依赖于 SQLiteOpenHelper 类。
- SQLite 数据库的基本操作依赖于 SQLiteDatabase 类。
- 可以使用原始的 SQL 语句操作 SQLite 数据库,也可以使用特定的方法来操作 SQLite 数据库。
- 为了保证数据的安全性,在数据库中引入了事务。

习题 6

1. Android 中可用的数据存储方式有_____、_____、_____和_____。
2. SharedPreferences 存储的是_____的数据。

3. 使用 SQLite 数据库时，创建数据库以及数据库版本更新需要继承_____。
4. 简述 Android 数据存储的方法。
5. 简述事务的四个基本要素。
6. 编写用户登录程序，要求以两种不同的方式保存和显示用户的信息。
7. 编写一个记事本程序，实现在界面中以列表的形式显示记录的信息，并能够对记录的信息进行增加、删除、修改和查询。

第7章 使用内容提供者共享数据

 **本章导图**

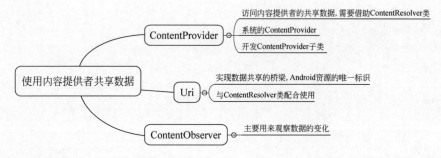

主要内容

- 内容提供者的创建及使用。
- 使用 ContentResolver 操作数据。
- 开发 ContentProvider 的子类。
- 使用 ContentResolver 调用方法。
- 系统的 ContentProvider。
- 监听 ContentProvider 的数据改变。

 难点

- 内容提供者的创建及使用。
- 使用 ContentResolver 操作数据。
- 监听 ContentProvider 的数据改变。

在第 6 章数据存储与 I/O 中学习了 SharedPreferences 存储、File 存储以及 SQLite 存储,但是这些存储的技术仅限于在当前的程序中使用。但是在 Android 开发中,有时也会访问其他应用程序的数据,这该怎么办呢? 为了解决这个问题,Android 系统为开发者提供了另外一种数据存储的技术——ContentProvider(内容提供者)。本章将针对内容提供者进行详细的讲解。

7.1 数据共享标准:ContentProvider

为什么要将自己程序中的数据共享给其他程序呢? 当然,这是要视情况而定的,例如账号和密码这样的隐私数据显然是不能共享给其他程序的,不过一些可以让其他程序进行二次开发的基础性数据,还是可以选择将其共享的。例如系统的电话簿程序,它的数据库中保存了很多联系人信息,如果这些数据都不允许第三方的程序进行访问,那么很多应用的功能都要大打折扣了。除了电话簿之外,还有短信、媒体库等程序都实现了跨程序数据共享的功能。ContentProvider 提供了在应用程序之前共享数据的一种机制,接下来就对这一技术进行深入的学习。

7.1.1 ContentProvider 简介

ContentProvider 是不同应用程序之间进行数据交换的标准 API, ContentProvider 以某种 Uri 的形式对外提供数据,允许其他应用访问或修改数据;其他应用程序使用 ContentResolver 根据 Uri 去访问操作指定数据。利用 ContentProvider 共享数据的工作原理如图 7.1 所示。

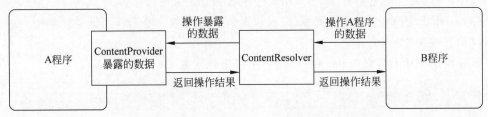

图 7.1 利用 ContentProvider 共享数据的工作原理

在图 7.1 中,A 程序需要使用 ContentProvider 暴露数据,才能被其他程序操作。B 程序必须通过 ContentResolver 操作 A 程序暴露出来的数据,而 A 程序会将操作结果返回给 ContentResolver,然后 ContentResolver 再将操作结果返回给 B 程序。然而实际上对于 ContentProvider 而言,无论数据的来源是什么,它都认为是一种表,然后把数据组织成表,以数据库的形式操作数据。例如系统中的联系人数据,联系人的信息可能会以表 7.1 所示的形式进行显示。

表 7.1 联系人数据

_ID	NAME	NUMBER	EMAIL
1	张子涵	123*****344	164**@qq.com
2	赵飞	135*****346	234**@126.com
3	王玉龙	183*****194	187**@qq.com
4	孙现伟	153*****917	476**@126.com

在表 7.1 中,每一行代表一条记录,每一列代表特定类型和含义的数据,每条记录包含一个数值类型的_ID 字段,主要是标识记录的唯一性,也可以根据该字段查询相关表格的数据。例如,在该表中可以根据_ID 查询联系人电话,在另外一个表中也可以根据该_ID 查询相关的短信信息。

如果需要得到表 7.1 中的任意一个字段,就需要知道每个字段所使用的数据类型,这样就可以使用 Cursor 提供的相同数据类型的方法,如 getInt()、getString()、getLong()等。

7.1.2 Uri 简介

ContentProvider 是允许不同应用进行数据交换的,ContentProvider 以 Uri 的形式提供数据的访问操作接口,而其他应用则通过 ContentResolver 去访问指定的数据,外界进程通过 Uri 找到对应的 ContentProvider 其中的数据,再进行数据操作。Uri 主要由三部分组成,分别是 scheme、authority 和 path,如图 7.2 所示。

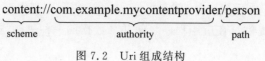

图 7.2 Uri 组成结构

Uri 组成部分说明如下:

(1) scheme 部分以 content://开头的标识,是 ContentProvider 的通用标准前缀,表明这个数据被 ContentProvider 所控制,它不会被修改。

(2) authority 部分表示为内容提供者设置的唯一标识,com.example.mycontentprovider 是在创建内容提供者时指定的 authorities 属性值,主要用来区分不同的应用程序,一般为了避免冲突,都会采用程序包名的方式来进行命名。

(3) path 部分/person 代表资源(或者数据),当访问者需要操作不同的资料时,这部分是可以动态改变的。

需要使用 Uri 工具类的 parse()方法将字符串转换为 Uri 的代码如下:

```
Uri uri = Uri.parse("content://com.example.mycontentprovider/person");
```

每个 ContentProvider 都会有一个 Uri,当对 ContentProvider 中的数据进行操作时,会通过对应的 Uri 指定相关的数据并进行操作。如果一个 ContentProvider 中含有多个数据源(例如多个表)时,就需要对不同的 Uri 进行区分,此时可以用 UriMatcher 类对 Uri 进行匹配,匹配步骤如下:

(1) 初始化 UriMatcher 类。

在 ContentProvider 中对 UriMatcher 类进行初始化。示例代码如下:

```
UriMatchermatcher = new UriMatcher(UriMatcher.NO_MATCH);
```

上述代码中,构造函数 UriMatcher()的参数表示 Uri 没有匹配成功的匹配码,该匹配码通常使用-1 来表示。在此处构造函数 UriMatcher()的参数可以设置为-1,也可以设置为 UriMatcher.NOMATCH,UriMatcher.NOMATCH 是一个值为-1 的常量。

(2) 注册需要的 Uri。

将需要用的 Uri 通过 addURI()方法注册到 UriMatcher 对象中。示例代码如下:

```
    matcher.addURI("com.example.contentprovider","people",PEOPLE);
    matcher.addURI("com.example.contentprovider","person/#",PEOPLE_ID);
```

上述代码中,addURI()方法中的第 1 个参数表示 Uri 的 authority 部分,第 2 个参数表示 Uri 的 path 部分,第 3 个参数表示 Uri 匹配成功后返回的匹配码。

(3) 与已经注册的 Uri 进行匹配。

在 ContentProvider 重写的 query()、insert()、update()、delete()方法中,可以通过 UriMatcher 对象的 match()方法来匹配 Uri,通过 switch 循环语句将每个匹配结果区分开,并做相应的操作。示例代码如下:

```
Uri uri = Uri.parse("content://" + "com.example.contentprovider" + "/people");
int match = matcher.match(uri);
switch (match){
    case PEOPLE:
    //匹配成功后做的相关操作
    case PEOPLE_ID:
    //匹配成功后做的相关操作
    default:
    return null;
}
```

7.1.3 使用 ContentResolver 操作数据

在不同的程序之间交换数据时,应用程序会通过 ContentProvider 暴露自己的数据,如果要访问 ContentProvider 中共享的数据,就需要借助 ContentResolver(内容解析器)类,通过 Context 中的 getContentResolver() 方法获取该类的实例。这个类提供了 insert()、update()、delete()、query() 方法用于添加、更新、删除和查询数据。这些方法和 SQLiteDatabase 中比较相似,但也有不同,这些方法不需要传递表名参数,需要用一个 Uri 参数,就可以清楚地表达出是哪个程序的哪张表。

1. 获取 ContentResolver 对象

示例代码如下:

```
ContentResolver resolver = getContentResolver();
```

2. 操作数据

前面也提到了 ContentProvider 是以类似数据库中表的方式将数据暴露出去,那么 ContentResolver 也将采用类似数据库的操作来从 ContentProvider 中获取数据。常用的方法如下。

(1) public Uri insert(Uri uri, ContentValues values):该方法用于向 ContentProvider 中添加数据。

(2) public int delete(Uri uri, String selection, String[] selectionArgs):该方法用于从 ContentProvider 中删除某些数据。

(3) public Cursor query(Uri uri, String[] projection, String selection, String[] selectionArgs, String sortOrder):该方法用于从 ContentProvider 中获取数据。

(4) public int update(Uri uri, ContentValues values, String selection, String[] selectionArgs):该方法用于修改 ContentProvider 中的某些信息。

使用 ContentResolver 访问 ContentProvider 共享数据的示例代码如下:

```
Uri uri = Uri.parse("content://com.example.mycontentprovider/");
ContentResolver resolver = getContentResolver();
Cursor cursor = resolver.query(uri,new String[]{"id","name","age","phone","address"},
null,null,null);
while(cursor.moveToNext()){
```

```
        int id = cursor.getInt(0);
        String name = cursor.getString(1);
        int age = cursor.getInt(2);
        String phone = cursor.getString(3);
        String address = cursor.getString(4);
    }
    cursor.close();
```

7.2 开发 ContentProvider

ContentProvider 是 Android 核心组件之中的一个,如果应用程序想要共享其数据,就需要定义一个类继承 ContentProvider 类,并重写该类用于提供数据和存储数据的方法,这样就可以向其他应用共享其数据。

7.2.1 开发 ContentProvider 的子类

如果要创建一个内容提供者,首先需要创建一个类继承抽象类 ContentProvider,例如创建一个名为 MyContentProvider 的应用程序,指定包名为 com.example.contentprovider。在程序包名处右击,选择 New→Other→ContentProvider 选项,在弹出的对话框中输入内容提供者的 Class Name(名称)和 URI Authorities(唯一标识,通常使用包名)。填写完成后点击 Finish 按钮,内容提供者便创建完成,如图 7.3 所示。

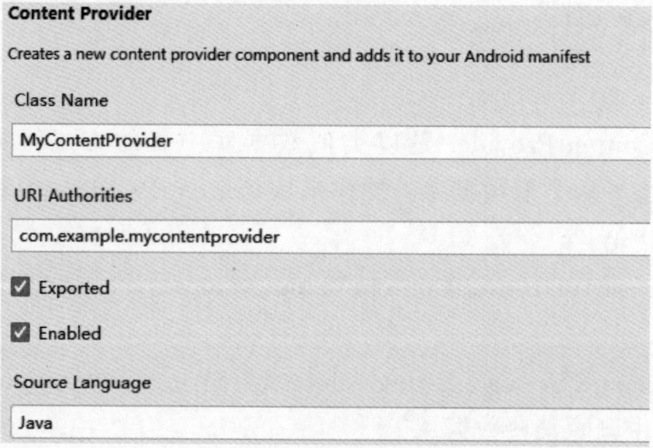

Content Provider
Creates a new content provider component and adds it to your Android manifest

Class Name
`MyContentProvider`

URI Authorities
`com.example.mycontentprovider`

☑ Exported
☑ Enabled

Source Language
`Java`

图 7.3 创建内容提供者

上述步骤完成后,打开 MyContentProvider.java,具体代码如下:

```
package com.example.mycontentprovider;

import android.content.ContentProvider;
import android.content.ContentValues;
import android.database.Cursor;
import android.net.Uri;

public class MyContentProvider extends ContentProvider {
    public MyContentProvider() {
```

```java
    }

    @Override
    public int delete(Uri uri, String selection, String[] selectionArgs) {
        throw new UnsupportedOperationException("Not yet implemented");
    }

    @Override
    public String getType(Uri uri) {
        throw new UnsupportedOperationException("Not yet implemented");
    }

    @Override
    public Uri insert(Uri uri, ContentValues values) {
        throw new UnsupportedOperationException("Not yet implemented");
    }

    @Override
    public boolean onCreate() {
        return false;
    }

    @Override
    public Cursor query(Uri uri, String[] projection, String selection,
                        String[] selectionArgs, String sortOrder) {
        throw new UnsupportedOperationException("Not yet implemented");
    }

    @Override
    public int update(Uri uri, ContentValues values, String selection,
                      String[] selectionArgs) {
        throw new UnsupportedOperationException("Not yet implemented");
    }
}
```

在上述代码中，可以看到重写了该类的 onCreate()、getType()、insert()、delete()、update()、query()方法。其中，onCreate()方法是在创建内容提供者时调用的；getType()方法用于返回指定 Uri 代表的数据类型 MIME，例如 Windows 系统中.txt 文件和.jpg 文件就是两种不同的 MIME 类型。insert()、delete()、update()、query()方法分别用于根据指定的 Uri 对数据进行增、删、改、查操作。

内容提供者创建完成后，Android Studio 会自动在 AndroidMainfest.xml 文件中对内容提供者进行注册，具体代码如下：

```xml
<application ……>
        ……
        <provider
            android:name=".MyContentProvider"
            android:authorities="com.example.mycontentprovider"
            android:enabled="true"
            android:exported="true" >
        </provider>
</application>
```

上述代码中，provider 标签中的配置用于注册创建的 MyContentProvider，该标签中设置的属性信息如下。

name：该属性的值是 MyContentProvider 的全名称（包含包名）。在 AndroidManifest.xml 文件中 MyContentProvider 的全名称可以用.MyContentProvider 来代替。

authorities：该属性的值标识了 MyContentProvider 提供的数据。该值可以是一个或多个 URI authority，多个 authority 名称之间需要用分号隔开，该属性的值通常设置为包名。

enabled：该属性的值表示 MyContentProvider 能否被系统实例化。如果属性 enabled 的值为 true，则表示可以被系统实例化；如果为 false，则表示不允许被系统实例化。该属性默认值为 true。

exported：该属性的值表示 MyContentProvider 能否被其他应用程序使用。如果属性 exported 的值为 true，则表示任何应用程序都可以通过 URI 访问 MyContentProvider；如果为 false，则表示只有用户 ID（程序的 build.gradle 文件中的 applicationId，applicationId 是每个应用的唯一标识）相同的应用程序才能访问到它。

需要注意的是，每个应用程序中创建的 ContentProvider 都必须在 AndroidManifest.xml 文件的 provider 标签中定义，否则，系统将找不到需要运行的 ContentProvider。

7.2.2 使用 ContentResolver 调用方法

ContentProvider 中重写的方法需要由调用者 ContentResolver 来调用才能触发，ContentResolver 传入 Uri 参数来调用这些方法。Context 提供了 getContentResolver()方法，因此 Activity、Service 等组件都可以通过 getContentResolver()方法获取 ContentResolver 对象。ContentResolver 类提供了与 ContentProvider 类相同签名的 4 个方法。示例代码如下：

```java
package com.example.mycontentprovider;
import androidx.appcompat.app.AppCompatActivity;
import android.content.ContentResolver;
import android.content.ContentValues;
import android.database.Cursor;
import android.net.Uri;
import android.os.Bundle;
import android.widget.Toast;

public class MainActivity extends AppCompatActivity {
    private ContentResolver resolver;
    private Uri uri = Uri.parse("content://com.example.mycontentprovider/");
    @Override
    protected void onCreate(Bundle savedInstanceState) {
        super.onCreate(savedInstanceState);
        setContentView(R.layout.activity_main);
        //获取系统的 ContentResolver 对象
        resolver = getContentResolver();
    }
    public void query(){
        //调用 ContentResolver 的 query()方法
        //实际返回的是该 Uri 对应的 ContentProvider 的 query()的返回值
```

```java
        Cursor cursor = resolver.query(uri,null,"query_where",null,null);
        Toast.makeText(this,
                "ContentResolver 返回的 Cursor 为:" + cursor,Toast.LENGTH_SHORT).
show();
    }
    public void insert(){
        ContentValues values = new ContentValues();
        values.put("name","AndroidStudio");
        //调用 ContentResolver 的 insert()方法
        //实际返回的是该 Uri 对应的 ContentProvider 的 insert()方法的返回值
        Uri newUri = resolver.insert(uri,values);
        Toast.makeText(this,
                "ContentResolver 新插入的记录的 Uri 为:" + newUri,Toast.LENGTH_SHORT).
show();
    }
    public void update(){
        ContentValues values = new ContentValues();
        values.put("name","Android");
        //调用 ContentResolver 的 update()方法
        //实际返回的是该 Uri 对应的 ContentProvider 的 update()方法的返回值
        int count = resolver.update(uri, values, "update_where", null);
        Toast.makeText(this,"ContentResolver 更新记录数为:" + count,Toast.LENGTH_SHORT).
show();
    }
    public void delete(){
        //调用 ContentResolver 的 delete()方法
        //实际返回的是该 Uri 对应的 ContentProvider 的 delete()方法的返回值
        int count = resolver.delete(uri, "delete_where", null);
        Toast.makeText(this,"ContentResolver 删除记录数为:" + count,Toast.LENGTH_SHORT).
show();
    }
}
```

在上述代码中,创建了 4 个方法,分别在这 4 个方法中使用 ContentResolver 对象的 query()、insert()、update()、delete()方法实现对应用数据的操作。

7.3 系统的 ContentProvider

Android 系统提供许多可以直接使用的系统 ContentProvider,例如:
(1) MediaProvider:用来查询磁盘上多媒体文件。
(2) ContactsProvider:用来查询联系人信息。
(3) CalendarProvider:用来提供日历相关信息的查询。
(4) BookmarkProvider:用来提供书签信息的查询。

这些 ContentProvider 的使用大同小异,使用它们对应的 Uri 地址就可以进行增、删、改、查的操作。Android 官方文档中提供相应 ContentProvider 的 Uri 和 ContentProvider 操作的数据列的列名,可以根据需要查阅 Android 官方文档。

本节以一个访问 MediaProvider 程序为例,展示如何使用系统提供的

ContentProvider。MediaProvider 作为系统级别的应用程序在系统上运行,专门负责收集多媒体文件(音频、视频、文件)相关的信息。当需要获取这类文件相关的信息时,可以向 MediaProvider 发起查询的请求。MediaProvider 在开机启动后,会在后台"监听"磁盘上文件的变化,特定情况下,会自动更新媒体文件的信息,例如磁盘上是否增加、修改或删除媒体文件等。

首先,需要确定访问的 Uri,Android 为多媒体提供了如下 Uri:

(1)MediaStore.Audio.Media.EXTERNAL_CONTENT_URI:存储在外部设备的音频文件。

(2)MediaStore.Audio.Media.INTERNAL_CONTENT_URI:存储在手机内部的音频文件。

(3)MediaStore.Images.Media.EXTERNAL_CONTENT_URI:存储在外部设备的图片文件。

(4)MediaStore.Images.Media.INTERNAL_CONTENT_URI:存储在内部设备的图片文件。

(5)MediaStore.Video.Media.EXTERNAL_CONTENTURI:存储在外部设备的视频文件。

(6)MediaStore.Video.Media.INTERNAL_CONTENT URI:存储在内部设备的视频文件。

我们创建的应用程序的主要功能是查看当前手机外部存储中所有图片的信息,因此查询请求的 Uri 地址为:

```
Uri uri = MediaStore.Images.Media.EXTERNAL_CONTENT_URI
```

例 7.1 GetImagesInfoApp7_1

由于想获取图片的一些基本信息,因此在布局的设计上,采用一个 TextView 控件用来显示获取的信息,如图 7.4 所示。

对应的 activity_main.xml 文件的代码如下所示:

```xml
<?xml version = "1.0" encoding = "utf-8"?>
<LinearLayout xmlns:android = "http://schemas.android.com/apk/res/android"
    xmlns:app = "http://schemas.android.com/apk/res-auto"
    xmlns:tools = "http://schemas.android.com/tools"
    android:layout_width = "match_parent"
    android:layout_height = "match_parent"
    tools:context = ".MainActivity"
    android:orientation = "vertical">
    <TextView
        android:id = "@+id/text_msg"
        android:layout_width = "match_parent"
        android:layout_height = "wrap_content"
        android:text = "TextView"
        android:textSize = "16sp"
        android:textStyle = "bold" />
</LinearLayout>
```

图 7.4 主界面布局

在前面知识讲解中曾提到,在 Android 6.0 以后,如果访问系统中存储的文件,不仅需要静态授权还需要动态授权,获取手机中图片信息的操作也是如此。第一步需要在 AndroidMainfest.xml 中进行授权,核心代码如下:

```
< uses - permission android:name = "android.permission.READ_EXTERNAL_STORAGE"/>
```

第二步需要进行动态授权,仍然采用第 6 章讲解的知识,首先引入需要的工具类 PermissionActivity.java 和 PermissionListener.java;接着将 MainActivity 继承的类修改为 PermissionActivity,这样就可以使用其中的请求授权方法了。

接着,编写交互的 MainActivity 文件,其对应的代码如下所示:

```java
package com.example.getimagesinfoapp7_1;
import android.Manifest;
import android.content.ContentResolver;
import android.database.Cursor;
import android.os.Bundle;
import android.provider.MediaStore;
import android.util.Log;
import android.provider.MediaStore.Images;
import android.widget.TextView;
import java.util.List;
public class MainActivity extends PermissionActivity {
    private boolean isGranted = false;
    private StringBuilder sb = new StringBuilder();
    private TextView textMsg;
    @Override
    protected void onCreate(Bundle savedInstanceState) {
        super.onCreate(savedInstanceState);
```

```java
        setContentView(R.layout.activity_main);
        textMsg = findViewById(R.id.text_msg);
        getImagesInfo();
        textMsg.setText(sb.toString());

    }
    private void getImagesInfo() {
        requestPermission();
        if (isGranted){
            ContentResolver resolver = getContentResolver();
            String[] projection = {
                    Images.Media._ID,
                    Images.Media.DATA,
                    Images.Media.WIDTH,
                    Images.Media.HEIGHT,
                    Images.Media.SIZE
            };
            Cursor c = resolver.query(MediaStore.Images.Media.EXTERNAL_CONTENT_URI,
                    projection, null, null, null);
            while(c.moveToNext()){
                int id = c.getInt(c.getColumnIndex(Images.Media._ID)); //通过列的索引得
                                                                       //到 ID 的值
                String path = c.getString(c.getColumnIndex(Images.Media.DATA));
                double width = c.getDouble(c.getColumnIndex(Images.Media.WIDTH));
                double height = c.getDouble(c.getColumnIndex(Images.Media.HEIGHT));
                double size = c.getDouble(c.getColumnIndex(Images.Media.SIZE));
                sb.append("id = ").append(id)
                        .append(",path = ").append(path)
                        .append(",width = ").append(width)
                        .append(",height = ").append(height)
                        .append(",size = ").append(size)
                        .append("\n");
            }
        }
    }
    //请求授权的方法
    private void requestPermission()
    {
        String[] permissionList = new String[]{Manifest.permission.READ_EXTERNAL_STORAGE};
        requestRunTimePermission(permissionList, new PermissionListener() {
            @Override
            public void onGranted() {
                isGranted = true;
            }

            @Override
            public void onGranted(List<String> grantedPermission) {

            }

            @Override
            public void onDenied(List<String> deniedPermission) {
                isGranted = false;
            }
        });
    }
}
```

在上述代码中,获取图片的信息使用了 Images.Media 常量:

(1) Images.Media.ID:图片的 ID 值,该值由系统创建。

(2) Images.Media.DISPLAY_NAME:图片的显示名。

(3) Images.Media.DESCRIPTION:图片的详细描述。

(4) Images.Media.DATA:图片的保存位置。

(5) Images.Media.TITLE:图片的标题。

(6) Images.Media.MIME_TYPE:图片的类型,格式为类型/子类型。

(7) Images.Media.SIZE:图片占用的空间,单位为字节。

(8) Images.Media.WIDTH:图片宽度。

(9) Images.Media.HEIGHT:图片的高度。

此外,在 MainActivity 中还定义了一个获取图片信息的方法,像查询数据库操作一样,将查询的信息存放到一个 StringBuilder 对象中,最后将其转换为字符串显示到 TextView 控件上,其运行结果如图 7.5 所示。

图 7.5 GetImagesInfoApp7_1 运行结果

7.4 监听 ContentProvider 的数据改变

通过前面的讲解可知,使用 ContentResolver 操作 ContentProvider 共享出来的数据。如果应用程序需要实时监听 ContentProvider 共享的数据是否发生变化,则需要使用 Android 系统提供的内容观察者 ContentObserver。本节将针对内容观察者 ContentObserver 进行详细的讲解。

内容观察者 ContentObserver 是用来观察指定 Uri 所代表的数据，当观察到指定 Uri 代表的数据发生变化时，就会触发 ContentObserver 的 onChange()方法。此时，在 onChange()方法里使用 ContentResovler 可以查询到变化的数据，其工作原理如图 7.6 所示。

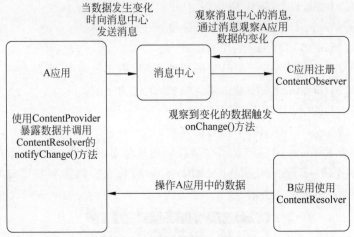

图 7.6　内容观察者工作原理

ContentObserver 中两个常用的方法如下：

public void ContentObserver(Handler handler)：ContentObserver 的派生类都需要调用该构造方法。参数可以是主线程 Handler，也可以是任何 Handler 对象（Handler 将在后续章节中进行讲解）。

public void onChange(boolean selfChange)：当观察到的 Uri 代表的数据发生变化时，会触发该方法。

ContentProvider 通过 delete()、insert()、update()方法使数据发生变化，因此使用 ContentObserver 同样需要观察这三个方法，并且需要在 ContentProvider 的这三个方法中调用 ContentResolver 的 notifyChange()方法。为了对内容观察者相关知识有一定的掌握，接下来就通过检测数据变化的示例来讲解如何使用内容观察者。由于该示例是检测数据库中数据的变化，因此在该示例中也会使用数据库的相关知识。

例 7.2　ContentObserverDBApp7_2

首先创建一个名为 ContentObserverDBApp7_2 的应用程序，将 res/layout 文件夹中的约束布局修改为垂直的线性布局，放置 4 个 Button 控件，分别用于显示"添加""修改""删除""查询"按钮，如图 7.7 所示。

其对应的详细代码如下所示：

图 7.7　操作数据库界面

```
<?xml version = "1.0" encoding = "utf - 8"?>
< LinearLayout xmlns:android = "http://schemas.android.com/apk/res/android"
    xmlns:app = "http://schemas.android.com/apk/res - auto"
```

```xml
xmlns:tools = "http://schemas.android.com/tools"
android:layout_width = "match_parent"
android:layout_height = "match_parent"
tools:context = ".MainActivity"
android:orientation = "vertical">

< Button
    android:id = "@ + id/btn_insert"
    android:layout_width = "match_parent"
    android:layout_height = "wrap_content"
    android:text = "添加"
    android:textSize = "16sp"
    android:textStyle = "bold"/>

< Button
    android:id = "@ + id/btn_update"
    android:layout_width = "match_parent"
    android:layout_height = "wrap_content"
    android:text = "修改"
    android:textSize = "16sp"
    android:textStyle = "bold"/>

< Button
    android:id = "@ + id/btn_delete"
    android:layout_width = "match_parent"
    android:layout_height = "wrap_content"
    android:text = "删除"
    android:textSize = "16sp"
    android:textStyle = "bold"/>

< Button
    android:id = "@ + id/btn_query"
    android:layout_width = "match_parent"
    android:layout_height = "wrap_content"
    android:text = "查询"
    android:textSize = "16sp"
    android:textStyle = "bold" />
</LinearLayout >
```

在该示例中用到了数据库，因此定义了一个类 PersonDBOpenHelper 继承自 SQLiteOpenHelper 类，在该类中创建数据库 person 和数据表 info，核心代码如下：

```java
package com.example.contentobserverdbapp7_2;
import android.content.Context;
import android.database.sqlite.SQLiteDatabase;
import android.database.sqlite.SQLiteOpenHelper;
import androidx.annotation.Nullable;
public class PersonDBOpenHelper extends SQLiteOpenHelper {
    private Context context;
    public PersonDBOpenHelper(@Nullable Context context) {
        super(context, "person", null, 1);
    }
```

```java
    @Override
    public void onCreate(SQLiteDatabase db) {
        db.execSQL("create table info (_id integer primary key autoincrement, name varchar(20))");
    }

    @Override
    public void onUpgrade(SQLiteDatabase sqLiteDatabase, int i, int i1) {

    }
}
```

接下来创建一个内容提供者 PersonContentProvider 继承自 ContentProvider, 在该类中实现对数据库中的数据进行添加、删除、修改、查询操作,核心代码如下:

```java
package com.example.contentobserverdbapp7_2;
import android.content.ContentProvider;
import android.content.ContentUris;
import android.content.ContentValues;
import android.content.UriMatcher;
import android.database.Cursor;
import android.database.sqlite.SQLiteDatabase;
import android.net.Uri;
public class PersonContentProvider extends ContentProvider {
    //定义一个 Uri 路径的匹配器,如果路径匹配不成功则返回-1
    private static UriMatcher mUriMatcher = new UriMatcher(-1);
    //匹配路径成功时的返回码
    private static final int SUCCESS = 1;
    //数据库操作类的对象
    private PersonDBOpenHelper helper;
    //添加路径匹配器的规则
    static {
        mUriMatcher.addURI("com.example.contentobserverdb", "info", SUCCESS);
    }
    //当内容提供者被创建时调用
    public boolean onCreate() {
        helper = new PersonDBOpenHelper(getContext());
        return false;
    }
    //查询数据操作
    public Cursor query(Uri uri, String[] projection, String selection,
                        String[] selectionArgs, String sortOrder) {
        //匹配查询的 Uri 路径
        int code = mUriMatcher.match(uri);
        if (code == SUCCESS) {
            SQLiteDatabase db = helper.getReadableDatabase();
            return db.query("info", projection, selection, selectionArgs,
                    null, null, sortOrder);
        } else {
            throw new IllegalArgumentException("路径不正确!");
        }
```

```java
}
//添加数据操作
public Uri insert(Uri uri, ContentValues values) {
    int code = mUriMatcher.match(uri);
    if (code == SUCCESS) {
        SQLiteDatabase db = helper.getReadableDatabase();
        long rowId = db.insert("info", null, values);
        if (rowId > 0) {
            Uri insertedUri = ContentUris.withAppendedId(uri, rowId);
            //提示数据库的内容变化了
            getContext().getContentResolver().notifyChange(insertedUri, null);
            return insertedUri;
        }
        db.close();
        return uri;
    } else {
        throw new IllegalArgumentException("路径不正确!");
    }
}
//删除数据操作
public int delete(Uri uri, String selection, String[] selectionArgs) {
    int code = mUriMatcher.match(uri);
    if (code == SUCCESS) {
        SQLiteDatabase db = helper.getWritableDatabase();
        int count = db.delete("info", selection, selectionArgs);
        //提示数据库的内容变化了
        if (count > 0) {
            getContext().getContentResolver().notifyChange(uri, null);
        }
        db.close();
        return count;
    } else {
        throw new IllegalArgumentException("路径不正确!");
    }
}
//更新数据操作
public int update(Uri uri, ContentValues values, String selection,
                String[] selectionArgs) {
    int code = mUriMatcher.match(uri);
    if (code == SUCCESS) {
        SQLiteDatabase db = helper.getWritableDatabase();
        int count = db.update("info", values, selection, selectionArgs);
        //提示数据库的内容变化了
        if (count > 0) {
            getContext().getContentResolver().notifyChange(uri, null);
        }
        db.close();
        return count;
    } else {
        throw new IllegalArgumentException("路径不正确!");
    }
}
```

```
    @Override
    public String getType(Uri uri) {
        return null;
    }
}
```

上述代码中,静态代码块主要是用来匹配 Uri 并且只会在程序启动时执行一次。代码块中通过 UriMatcher 的 addURI()方法添加需要匹配的 Uri,该方法中的第 1 个参数表示 Uri 的 authority 部分,第 2 个参数表示 Uri 的 path 部分,第 3 个参数表示 Uri 匹配成功的一个匹配码,在此处用一个常量 SUCCESS 来记录。

代码中还重写了 query()、insert()、update()、delete()方法,这些方法用于操作数据库中的数据,当 Uri 匹配成功时,便可以通过数据库类的 query()、insert()、update()、delete()方法来操作数据库表 info 中的信息,否则就会抛出不合法的参数异常 IllegalArgumentException。

最后编写交互的 MainActivity 类,该类中实现操作数据库界面上"添加"按钮、"修改"按钮、"删除"按钮以及"查询"按钮的点击事件,点击这 4 个按钮会分别调用 ContentResolver 对象的 insert()、update()、delete()、query()方法对数据库中的数据进行增、删、改、查的操作。核心代码如下所示:

```java
package com.example.contentobserverdbapp7_2;
import androidx.appcompat.app.AppCompatActivity;
import android.content.ContentResolver;
import android.content.ContentValues;
import android.database.ContentObserver;
import android.database.Cursor;
import android.database.sqlite.SQLiteDatabase;
import android.net.Uri;
import android.os.Bundle;
import android.os.Handler;
import android.util.Log;
import android.view.View;
import android.widget.Button;
import android.widget.Toast;
import java.util.ArrayList;
import java.util.HashMap;
import java.util.List;
import java.util.Map;
import java.util.Random;

public class MainActivity extends AppCompatActivity {
    private Button btnInsert, btnUpdate, btnDelete, btnQuery;
    private ContentResolver resolver;
    private Uri uri;

    @Override
    protected void onCreate(Bundle savedInstanceState) {
        super.onCreate(savedInstanceState);
        setContentView(R.layout.activity_main);
        initView();
        createDB();
```

```java
resolver = getContentResolver();
uri = Uri.parse("content://com.example.contentobserverdb/info");
resolver.registerContentObserver(uri, true, new MyObserver(new Handler()));
//添加事件
btnInsert.setOnClickListener(new View.OnClickListener() {
    @Override
    public void onClick(View view) {
        Random random = new Random();
        ContentValues values = new ContentValues();
        values.put("name", "android_content" + random.nextInt(10));
        Uri newuri = resolver.insert(uri, values);
        Toast.makeText(MainActivity.this, "添加成功", Toast.LENGTH_SHORT).show();
        Log.i("数据库应用:", "添加");
    }
});
//删除事件
btnDelete.setOnClickListener(new View.OnClickListener() {
    @Override
    public void onClick(View view) {
        //返回删除数据的条目数
        int deleteCount = resolver.delete(uri, "name = ?",
                new String[]{"android_content0"});
        Toast.makeText(MainActivity.this, "成功删除了" + deleteCount + "行",
                Toast.LENGTH_SHORT).show();
        Log.i("数据库应用:", "删除");
    }
});
//修改事件
btnUpdate.setOnClickListener(new View.OnClickListener() {
    @Override
    public void onClick(View view) {
        //将数据库 info 表中 name 为 android_content1 的这条记录修改为
        //update_android
        ContentValues values = new ContentValues();
        values.put("name", "update_android");
        int updateCount = resolver.update(uri, values, "name = ?",
                new String[]{"android_content1"});
        Toast.makeText(MainActivity.this, "成功修改了" + updateCount + "行",
                Toast.LENGTH_SHORT).show();
        Log.i("数据库应用:", "修改");
    }
});
//查询事件
btnQuery.setOnClickListener(new View.OnClickListener() {
    @Override
    public void onClick(View view) {
        List<Map<String, String>> data = new ArrayList<Map<String, String>>();
        //返回查询结果,是一个指向结果集的游标
        Cursor cursor = resolver.query(uri, new String[]{"_id", "name"},
                null, null, null);
        //遍历结果集中的数据,将每一条遍历的结果存储在一个 List 的集合中
        while (cursor.moveToNext()) {
            Map<String, String> map = new HashMap<String, String>();
```

```java
                    map.put("_id", cursor.getString(0));
                    map.put("name", cursor.getString(1));
                    data.add(map);
                }
                //关闭游标,释放资源
                cursor.close();
                Log.i("数据库应用:", "查询结果:" + data.toString());
            }
        });
    }

    //初始化界面
    private void initView() {
        btnInsert = findViewById(R.id.btn_insert);
        btnUpdate = findViewById(R.id.btn_update);
        btnDelete = findViewById(R.id.btn_delete);
        btnQuery = findViewById(R.id.btn_query);
    }

    //创建数据库
    private void createDB() {
        //创建数据库并向 info 表中添加 3 条数据
        PersonDBOpenHelper helper = new PersonDBOpenHelper(this);
        SQLiteDatabase db = helper.getWritableDatabase();
        for (int i = 0; i < 3; i++) {
            ContentValues values = new ContentValues();
            values.put("name", "android_content" + i);
            db.insert("info", null, values);
        }
        db.close();
    }

    //定义一个内容观察者类
    private class MyObserver extends ContentObserver {
        public MyObserver(Handler handler) {//handler 是一个消息处理器
            super(handler);
        }

        @Override
        //当 info 表中的数据发生变化时执行该方法
        public void onChange(boolean selfChange) {
            Log.i("监测数据变化", "数据库数据被改动!");
            super.onChange(selfChange);
        }
    }
}
```

在上述代码中,创建了一个 initView()方法用来初始化界面中的 Button 控件;创建了一个 createDB()方法用来初始化数据库及表;对 4 个按钮设置监听事件,实现界面上"添加"按钮、"修改"按钮、"删除"按钮、"查询"按钮的点击事件。在"添加"按钮的点击事件中,首先创建了一个产生随机数的对象 random,接着调用 ContentValues 对象的 put()方法,将要添加的数据放入 ContentValues 对象中。调用 ContentResolver 对象的 insert()方法将数据添加到数据库中，insert（）方法中的第 1 个参数表示该程序中内容提供者 PersonContentProvider 的 Uri，第 2 个参数表示要添加的数据，最后通过 Toast 与 Log 来提示添加成功的信息。界面上"修改"按钮、"删除"按钮、"查询"按钮的点击事件中分别通过 ContentResolver 对象的 update（）方法、delete（）方法以及 query（）方法来实现对数据库中的数据进行修改、删除以及查询的操作，在此不再进行详细解释。

在上述代码中，为了检测数据的变化，直接在 MainActivity 中定义一个 MyObserver 类继承自 ContentObserver 并使用 registerContentObserver（）方法进行注册。当数据发生变化时，就会执行 MyObserver 类中的 onChange（）方法。当执行上述程序点击按钮时，LogCat 打印的运行结果如图 7.8 所示。

图 7.8　ContentObserverDBApp7_2 运行结果

7.5　应用实例：读取联系人

观看视频

现在大多数 Android 应用程序都会读取用户的联系人、位置等权限，这是很普遍的。从功能上来讲，读取联系人可以判断用户的亲朋好友中有没有人使用同样的应用。例如微信的"添加朋友"里面有一个"手机联系人"，就需要用到手机通讯录的权限。通过这个权限可以看到用户的通讯录中有谁也在用微信，方便添加好友。接下来通过一个应用实例来演示如何使用 ContentProvider 来读取手机的通讯录。

例 7.3　ReadContactApp7_3

首先创建一个名为 ReadContactApp7_3 的应用程序，在 res/layout 文件夹中的布局文件 activity_main.xml 中放置显示列表信息的 ListView 控件，主界面如图 7.9 所示。

activity_main.xml 核心代码如下所示：

```xml
<?xml version="1.0" encoding="utf-8"?>
<LinearLayout xmlns:android="http://schemas.android.com/apk/res/android"
    xmlns:app="http://schemas.android.com/apk/res-auto"
    xmlns:tools="http://schemas.android.com/tools"
    android:layout_width="match_parent"
    android:layout_height="match_parent"
    tools:context=".MainActivity"
    android:orientation="vertical">

    <ListView
        android:id="@+id/list_view"
        android:layout_width="match_parent"
        android:layout_height="match_parent" />
</LinearLayout>
```

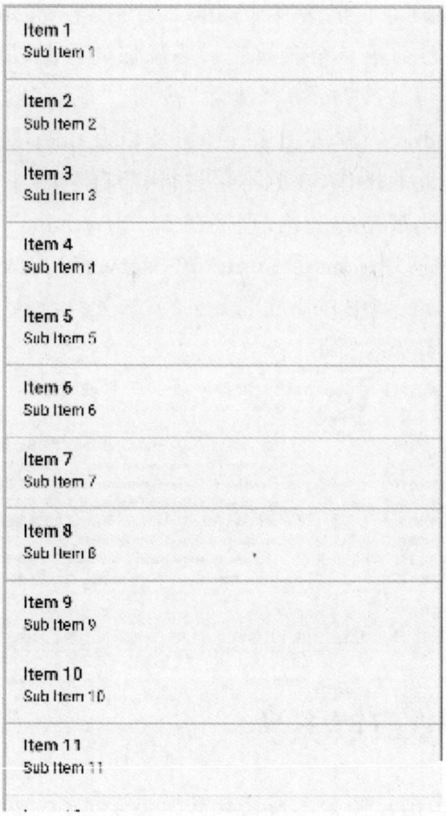

图 7.9 主界面

因为在主界面中使用了 ListView 控件,所以需要设置每一项的布局样式。在 res/layout 文件夹中新建一个相对布局,在布局中放置 4 个 TextView 控件用于显示联系人的姓名和电话,核心代码如下:

```xml
<?xml version = "1.0" encoding = "utf-8"?>
<RelativeLayout xmlns:android = "http://schemas.android.com/apk/res/android"
    android:layout_width = "match_parent"
    android:layout_height = "match_parent"
    android:padding = "6dp">

    <TextView
        android:id = "@+id/textView"
        android:layout_width = "wrap_content"
        android:layout_height = "wrap_content"
        android:layout_alignParentStart = "true"
        android:layout_alignParentTop = "true"
        android:text = "姓名:"
        android:textSize = "16sp"
        android:textStyle = "bold" />

    <TextView
        android:id = "@+id/tv_name"
        android:layout_width = "wrap_content"
```

```
            android:layout_height = "wrap_content"
            android:layout_toRightOf = "@ + id/textView"
            android:text = "张雨"
            android:textSize = "16sp"
            android:textStyle = "bold" />
    < TextView
            android:id = "@ + id/textView1"
            android:layout_width = "wrap_content"
            android:layout_height = "wrap_content"
            android:layout_below = "@ + id/textView"
            android:text = "电话:"
            android:textSize = "16sp"
            android:textStyle = "bold" />

    < TextView
            android:id = "@ + id/tv_phone"
            android:layout_width = "wrap_content"
            android:layout_height = "wrap_content"
            android:layout_below = "@ + id/textView"
            android:layout_toRightOf = "@ + id/textView1"
            android:text = "18236782195"
            android:textSize = "16sp"
            android:textStyle = "bold" />
</RelativeLayout >
```

接下来需要对读取联系人进行授权,不然无法读取手机中的通讯录,第一步在AndroidManifest.xml中进行静态授权,代码如下:

```
< uses - permission android:name = "android.permission.READ_CONTACTS"/>
```

第二步需要进行动态授权,仍然采用第 6 章讲解的知识,首先引入需要的工具类PermissionActivity.java 和 PermissionListener.java;接着将 MainActivity 继承的类修改为 PermissionActivity,这样就可以使用其中的请求授权方法了。

最后编写交互的 MainActivity,其核心代码如下所示:

```
package com.example.readcontactapp7_3;
import android.Manifest;
import android.database.Cursor;
import android.os.Bundle;
import android.provider.ContactsContract;
import android.widget.ListView;
import android.widget.SimpleAdapter;
import android.widget.Toast;
import java.util.ArrayList;
import java.util.HashMap;
import java.util.List;
import java.util.Map;

public class MainActivity extends PermissionActivity {
    private ListView listView;
    private boolean isGranted = false;
```

```java
            private List<Map<String,String>> listmaps = new ArrayList<>();
            private SimpleAdapter adapter;
            @Override
            protected void onCreate(Bundle savedInstanceState) {
                super.onCreate(savedInstanceState);
                setContentView(R.layout.activity_main);
                listView = findViewById(R.id.list_view);
                readContacts();
                adapter = new SimpleAdapter(
                        this,
                        listmaps,
                        R.layout.layout_item,
                        new String[]{"name","phone"},
                        new int[]{R.id.tv_name,R.id.tv_phone});
                listView.setAdapter(adapter);
                adapter.notifyDataSetChanged();

            }
            //读取联系人的方法
            private void readContacts()
            {
                requestPermission();
                if(isGranted == true) {
                    Cursor cursor = null;
                    try {
                        cursor = getContentResolver().query(ContactsContract.CommonDataKinds.Phone.
                            CONTENT_URI, null, null, null, null);
                        if (cursor != null) {
                            while (cursor.moveToNext()) {
                                String dispalyName = cursor.getString(cursor.getColumnIndex
(ContactsContract.
                                        CommonDataKinds.Phone.DISPLAY_NAME));
                                String phone = cursor.getString(cursor.getColumnIndex(ContactsContract.
                                        CommonDataKinds.Phone.NUMBER));
                                Map map = new HashMap();
                                map.put("name", dispalyName);
                                map.put("phone", phone);
                                listmaps.add(map);
                            }
                        }
                    } catch (Exception e) {
                        e.printStackTrace();
                    } finally {
                        if (cursor != null) {
                            cursor.close();
                        }
                    }
                }
                else
                {
                    Toast.makeText(this,
                        "You denied the EAD_CONTACTS Permission", Toast.LENGTH_SHORT).show();
```

```
        }
    }
    //请求授权的方法
    private void requestPermission()
    {
        String[] permissionList = new String[]{Manifest.permission.READ_CONTACTS};
        requestRunTimePermission(permissionList, new PermissionListener() {
            @Override
            public void onGranted() {
                isGranted = true;
            }
            @Override
            public void onGranted(List<String> grantedPermission) {

            }
            @Override
            public void onDenied(List<String> deniedPermission) {

            }
        });
    }
}
```

在上述代码中,定义了一个动态请求授权的方法 requestPermission()和一个读取通讯录的方法 readContacts()。在读取联系人信息之前,需要判断是否授权成功,授权成功之后才能进行读取操作,将读取的信息存储到 List<Map<String,String>>类型的集合中。创建一个 SimpleAdapter 类型的数据适配器对象,将集合中的数据显示到 ListView 控件上。程序运行结果如图 7.10 所示。

图 7.10　ReadContactApp7_3 运行结果

7.6 小结

- ContentProvider 为存储和读取数据提供了统一的接口。
- 应用继承 ContentProvider 类，并重写该类用于提供数据和存储数据的方法，就可以向其他应用共享其数据。
- 在 ContentProvider 中使用的查询字符串有别于标准的 SQL 查询，使用一种特殊的 Uri 来进行。
- 当外部应用需要对 ContentProvider 中的数据进行添加、删除、修改和查询操作时，可以使用 ContentResolver 类来完成。
- ContentObserver 目的是观察（捕捉）特定 Uri 引起的数据库的变化，继而做一些相应的处理。

习题 7

1. 在 ContentProvider 中 ContentUris 的作用是（　　）。
   A. 用于获取 Uri 路径后面的 ID 部分
   B. 增、删、改、查的方法都在这个类中
   C. 用于添加 Uri 的类
   D. 根本就用不到这个类，没关系
2. 利用内容解析者查询短信数据时，Uri 的编写正确的是（　　）。
   A. Uri uri = Uri.parse("content://sms");
   B. Uri uri = Uri.parse("content://sms/data");
   C. Uri uri = Uri.parse("content://sms/contact");
   D. Uri uri = Uri.parse("sms/");
3. Android 中创建内容提供者要继承（　　）。
   A. ContentData   B. ContentProvider
   C. ContentObserver   D. ContentDataProvider
4. 在 Android 中的 Activity 通过下面（　　）方法来得到 ContentResolver 的实例对象。
   A. new ContentResolver()   B. getContentResolver()
   C. newInstance()   D. ContentUris.newInstance()
5. 若要实现对系统联系人的增、删、改、查，需要使用的系统 ContentProvider 的 Uri 为（　　）。
   A. Contacts.Photos.CONTENT_URI
   B. Contacts.People.CONTENT_URI
   C. Contacts.Phones.CONTENT_URI
   D. Media.EXTERNAL_CONTENT_URI
6. 当观察到的 Uri 代表的数据发生变化时，会触发 ContentObserver 中的（　　）方法。

A. onCreate()　　　　　　　　　　B. notifyChange()
　　C. onChange()　　　　　　　　　　D. 以上说法都不对
7. 短信的内容提供者是(　　)。
　　A. ContactProvider　　　　　　　　B. MessageProvider
　　C. SmsProvider　　　　　　　　　　D. TelephonyProvider
8. 简述内容提供者的工作原理。
9. 简述内容观察者的工作原理。
10. 根据读取联系人的实例,编写一个程序读取手机短信并以列表的形式进行显示。

第8章 广播机制

 本章导图

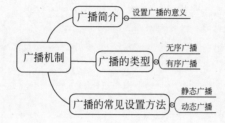

 主要内容

- 广播机制的基本概念。
- 广播的应用场景。
- 静态注册与动态注册的区别。
- 有序/无序广播的定义。
- 常用广播的定义。

 难点

- 广播的应用场景。
- 广播的动态注册。

在 Android 中,广播机制是一种用于应用程序内或应用程序之间传递消息和通知的机制。它允许应用程序的组件(如活动、服务或广播接收器)发送广播事件,而其他组件可以注册接收这些事件或通知,并采取相应的操作。

本章介绍 Android 系统中有关广播的基本概念,如广播的应用场景、广播的静态注册与动态注册、常用广播如电源充电状态的检测等,这些概念及方法在 Android 的开发场景中都是常见的进行性能提升、软件优化经常需要考虑的。

8.1 广播机制简介

基于 Android 系统的智能设备本身是一款功能强大的设备,在做硬件设计时,已经安装、配备了大量的具有不同功能的硬件组件。其中许多硬件组件以不同的方式影响着

Android 系统的运行状态,例如加速度计能够检测手机设备的当前物理方向(纵向或横向),Wi-Fi 能够发现新的可用网络,并在设备与网络的连接发生变化时对系统发出广播通知,而光传感器可以根据设备周边环境控制屏幕的亮度,设备上的硬件按钮可以触发中断,从而在系统中生成一些特定的广播事件。

除此之外,Android 系统总是会尽可能少地消耗电量,以使电池使用时间尽可能延长。然而,由于 Android 应用程序在控制设备的程度上具有相对较高的自由度,因此作为开发人员,对设备的不断变化的事件和更改做出反应是非常重要的;否则,它们可能会导致不必要的功率消耗。通过接收不同的广播事件,程序设计者可以使自己开发的应用程序更加健壮,并与整个 Android 系统协同工作。例如,如果程序设计者设计的软件不注重广播通知,使得非 Wi-Fi 环境下大量下载某些站点的原始图片、视频,导致用户流量快速耗尽。那么这个软件可能就不是一个能够广受欢迎的软件,如果这样的话,可能导致用户会寻求更好的替代软件。在流量控制方面,微信、今日头条做得就比较好,如微信可以选择是否下载原始图片、视频在用户点击之后才下载,而今日头条是在非 Wi-Fi 环境下,提示用户观看视频会消耗多少流量。

作为一名应用程序开发人员,我们可以通过 Android 的广播机制发送广播信息,这些广播信息可能是 Android 中预先由 API 定义的,也可以是由第三方应用程序定义,或者是由开发者自行定义。通过在自己的程序中监听或者发送某些广播信息,可以优化程序的后台调度,提高程序的性能,这样做在实际开发中是非常有用的。

Android 系统中的广播机制是基于观察者模式的,在相关代码的架构上分为广播发送者与广播接收者。其中,广播发送者是观察者(发布者),而广播接收者是订阅者。广播发送者将广播事件发布到系统中,然后系统将广播事件传递给已注册的接收者,以便它们可以采取相应的操作。当一个应用程序发送广播时,它将广播事件发送到系统,然后系统将广播事件传递给已注册的接收者。接收者可以是同一个应用程序内的组件,也可以是不同应用程序的组件。

在 Android 的广播机制中涉及以下一些重要的概念:

(1) 广播发送者(BroadcastSender)。广播发送者是负责发送广播的组件,通常是应用程序的一部分,如活动(Activity)或服务(Service)。广播发送者使用 Intent 对象来定义广播的类型和数据,并使用 sendBroadcast()或 sendOrderedBroadcast()方法发送广播。

(2) 广播接收者(BroadcastReceiver)。广播接收者是用于接收广播事件的组件。接收者可以在 AndroidManifest.xml 文件中声明,以在应用程序启动之前自动接收广播,也可以在运行时使用 registerReceiver()方法动态注册。接收者需要实现 BroadcastReceiver 类,并重写 onReceive()方法以处理接收到的广播事件。

(3) 广播类型。Android 提供了一些预定义的广播类型,例如系统广播(如屏幕开关机、进入和退出飞行模式、电池电量变化)、自定义广播(应用程序内部发送的事件)和有序广播(按照优先级依次传递给接收者)。

(4) 广播过滤器(IntentFilter)。广播过滤器用于指定接收者所感兴趣的广播类型。过滤器可以根据广播的动作(Action)、数据(Data)、类别(Category)等属性进行匹配。接收者可以在静态注册或动态注册时指定广播过滤器。

(5) 本地广播。本地广播是一种特殊类型的广播,仅在应用程序内部传递。与全局广播相比,本地广播更加高效,安全性更高,因为它不会向外部应用程序发送广播。本地广播可以使用 LocalBroadcastManager 类进行发送和接收广播。

(6)广播权限。某些广播需要特定的权限才能接收。广播发送者可以在 Intent 对象中设置权限,只有具有相应权限的接收者才能接收到该广播。

广播机制是 Android 应用程序中常用的一种通信方式,可用于发送应用内部的事件通知、接收系统级别的广播事件,以及与其他应用程序进行通信。使用广播,应用程序可以实现组件之间的松耦合通信,提高应用程序的灵活性和可扩展性。

8.2 发送广播

在 Android 中,发送广播的过程涉及以下步骤。

(1)创建 Intent 对象。首先,需要创建一个 Intent 对象,用于描述广播的类型和数据。Intent 对象可以包含一个动作来标识广播的目的或意图,也可以添加其他额外的信息,如数据、类别等。

(2)发送广播。使用广播发送方法将 Intent 对象发送出去。Android 提供了几种发送广播的方法,常用的包括:

- sendBroadcast(Intent):发送普通无序广播,将广播同时发送给所有匹配的接收者。
- sendOrderedBroadcast(Intent, receiverPermission):发送有序广播,按照优先级依次传递给接收者。
- sendStickyBroadcast(Intent):发送黏性广播,允许接收者在注册之前接收广播。

(3)广播传递。Android 系统接收到广播后,会将广播事件传递给已注册的接收者。系统会根据广播的类型和过滤器,找到匹配的接收者并调用其相应的广播接收器的 onReceive()方法。

(4)接收广播。广播接收者通过实现 BroadcastReceiver 类,并重写 onReceive()方法来处理接收到的广播事件。在 onReceive()方法中,可以根据 Intent 对象中的信息进行相应的操作。

需要注意的是,广播接收者可以通过静态注册或动态注册的方式进行注册。静态注册是在 AndroidManifest.xml 文件中声明广播接收者,使其在应用程序启动之前自动接收广播。动态注册是在运行时通过调用 registerReceiver()方法来注册广播接收者,可以更灵活地控制广播接收的时机。

通过发送广播,应用程序可以实现组件之间的通信和交互,例如发送应用内部的事件通知、接收系统级别的广播事件或与其他应用程序进行通信。广播机制为应用程序提供了一种松散耦合的通信方式,提高了应用程序的灵活性和可扩展性。

8.2.1 定义广播接收者

BroadcastReceiver 是 Android 中的一个基本组件,用于接收和处理广播消息。它充当了观察者模式中的接收者(订阅者),用于接收发送者(发布者)发送的广播事件。

首先在 Android Studio 中创建一个工程,这里命名为 MyBroadcastReceiverApp,包名为 com.example.mybroadreceiverapp,接着在程序的包名处(应用程序已经存在)右击,选择 New→Other→Broadcast Receiver 选项，如图 8.1 所示。

在下一步中，确定类名为 MyBroadcastReceiver、Exported 和 Enabled 选项已经勾选、语言为 Java 后，点击 Finish 按钮完成服务的创建，如图 8.2 所示。

根据图 8.1 和图 8.2 中的步骤创建完成的 MyBroadcastReceiver 的具体代码如【文件 8-1】所示。

【文件 8-1】 MyBroadcastReceiver.java

```java
package com.example.mybroadcastreceiverapp;
import android.content.BroadcastReceiver;
import android.content.Context;
import android.content.Intent;
public class MyBroadcastReceiver extends BroadcastReceiver {
    @Override
    public void onReceive(Context context, Intent intent) {
        throw new UnsupportedOperationException("Not yet implemented");
    }
}
```

MyBroadcastReceiver 继承自 BroadcastReceiver 类，重写了父类中的 onReceive()方法，由于此时 onReceive()方法还未实现，程序默认抛出一个未支持操作的异常 UnsupportedOperationException，在后续完善程序 onReceive()方法时，删除默认抛出的未支持操作的异常。

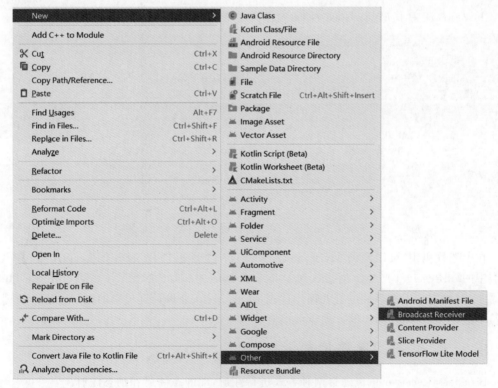

图 8.1　创建 Broadcast Receiver

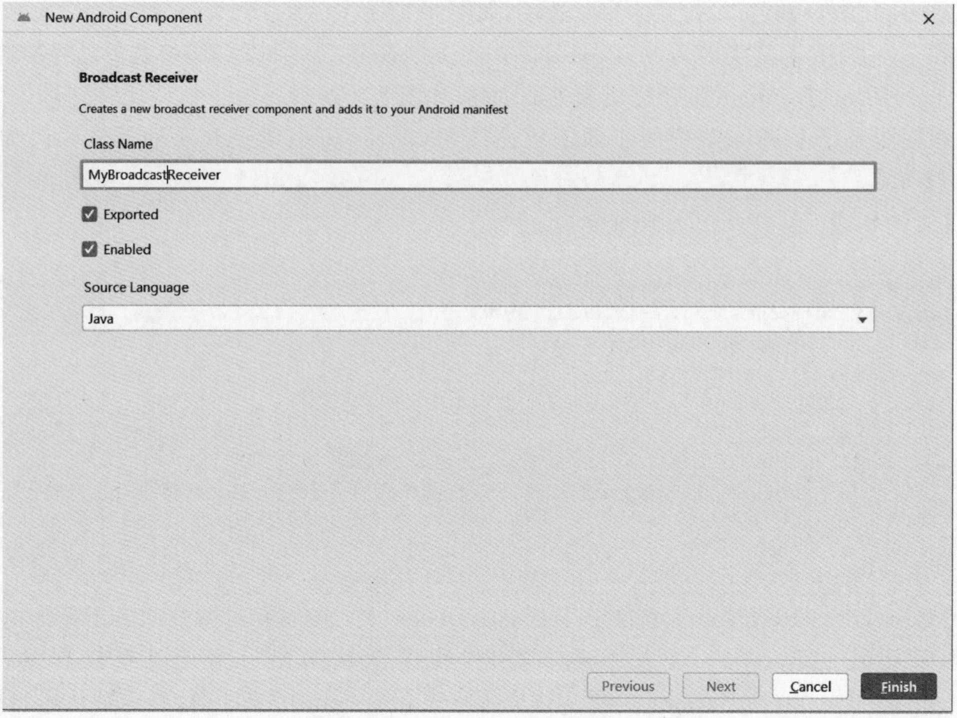

图 8.2　确认 BroadcastReceiver 的相关信息

下面程序完善了 MyBroadcastReceiver 类的定义，可以对指定的广播事件做出响应。

```java
public class MyBroadcastReceiver extends BroadcastReceiver {
    @Override
    public void onReceive(Context context, Intent intent) {
        //在接收到广播时执行的操作
        if (intent.getAction().equals("com.example.chapter08.MY_CUSTOM_ACTION"))
        {
        //处理自定义广播事件
        String message = intent.getStringExtra("message");
        Toast.makeText(context, "Received Broadcast: " + message,
                    Toast.LENGTH_SHORT).show();
        }
    }
}
```

在上述代码中，可以看到 MyBroadcastReceiver 类继承自 BroadcastReceiver 类，重写了父类中的 onReceive()方法，当接收到 com.example.chapter08.MY_CUSTOM_ACTION 广播时，对自定义广播事件进行处理。首先，通过 intent 获取传递过来的以 message 为键的字符串值，之后使用 Toast 的方式显示该值。

8.2.2　注册广播接收者

BroadcastReceiver 可以通过两种方式进行注册，即静态注册和动态注册。

1. 静态注册

静态注册需要在 AndroidManifest.xml 文件中声明对应的 BroadcastReceiver，并且指

定其接收的广播类型和过滤条件。这样，系统会在应用程序启动之前自动注册并在合适的时机调用相应的 BroadcastReceiver。

MyBroadReceiverApp 工程中的 AndroidManifest.xml 如【文件 8-2】所示。

【文件 8-2】 AndroidManifest.xml

```xml
<?xml version = "1.0" encoding = "utf-8"?>
<manifest xmlns:android = "http://schemas.android.com/apk/res/android"
    xmlns:tools = "http://schemas.android.com/tools">
    <application ……>
    ……
        <receiver
            android:name = ".MyBroadcastReceiver"
            android:enabled = "true"
            android:exported = "true">
            <intent-filter>
                <action android:name = "com.example.chapter08.MY_CUSTOM_ACTION" />
            </intent-filter>
        </receiver>
    </application>
</manifest>
```

在上述 AndroidManifest.xml 文件中，receiver 标签用于声明应用中的广播接收者组件，它定义了服务的属性和特性。Android:name 为指定的广播接收者名，这里是 MyBroadcastReceiver，由于该广播接收者代码和项目路径一致，因此前面的包名省略，仅保留一个点号"."。android:enable 和 android:exported 均为真，代表该广播接收者可以使用并可以被其他应用程序导出。

下面的 intent-filter 指定接收的广播动作为 com.example.chapter08.MY_CUSTOM_ACTION，也就是 MyBroadcastReceiver 这个广播接收者能够接收到的广播为 com.example.chapter08.MY_CUSTOM_ACTION。

2．动态注册

动态注册是使用代码编程的方式，在代码中使用 registerReceiver() 方法进行动态注册，这样可以在运行时灵活地控制 BroadcastReceiver 的注册和注销。

下面给出动态注册和注销广播接收者的示例程序。

```java
MyBroadcastReceiver receiver = new MyBroadcastReceiver();
IntentFilter filter = new IntentFilter("com.example.chapter08.MY_CUSTOM_ACTION");
//调用 registerReceiver()方法实现广播接收者的动态注册
registerReceiver(receiver, filter);
…
//调用 unregisterReceiver()方法实现广播接收者的动态注销
unregisterReceiver(receiver);
```

8.2.3 发送广播步骤

当完成广播接收者的定义和注册之后，就需要通过某些事件产生广播，使得广播接收者能够接收到对应的广播并做出响应。

MyBroadReceiverApp 工程中的 MainActivity.java 如【文件 8-3】所示。

【文件8-3】 MainActivity.java

```java
package com.example.mybroadcastreceiverapp;
……//省略导入的相关包、类
public class MainActivity extends AppCompatActivity implements View.OnClickListener{
    private MyBroadcastReceiver myBroadcastReceiver;
    private Button mButton;
    @Override
    protected void onCreate(Bundle savedInstanceState) {
        super.onCreate(savedInstanceState);
        Log.i("MSG","onCreate");
        setContentView(R.layout.activity_main);
        mButton = findViewById(R.id.btn_selfdefinebroadcast);
        mButton.setOnClickListener(this);
    }
    @Override
    public void onClick(View v) {
        //创建一个Intent对象,指定广播的动作
        //Intent intent = new Intent("com.example.chapter08.MY_CUSTOM_ACTION");
        Intent intent = new Intent();
        intent.setAction("com.example.chapter08.MY_CUSTOM_ACTION");
        //添加额外的数据到Intent对象
        intent.putExtra("message", "Hello, Broadcast1!");
        //发送广播
        sendBroadcast(intent);
        Log.i("MSG","发送自定义广播事件1");
    }
}
//创建一个Intent对象,指定广播的动作
Intent intent = new Intent("com.example.chapter08.MY_CUSTOM_ACTION");
//添加额外的数据到Intent对象
intent.putExtra("message", "Hello, Broadcast!");
//调用sendBroadcast()方法发送广播,注意需要设定好相关intent
sendBroadcast(intent);
```

MainActivity.java中使用findViewById()方法找到ID为btn_selfdefinebroadcast的Button按钮,并在对应的按钮点击事件中发送com.example.chapter08.MY_CUSTOM_ACTION广播。在以编程方式注册BroadcastReceiver时,必须在匹配的回调中注销它。在本例中,广播接收者是在onResume()方法中注册了com.example.chapter08.MY_CUSTOM_ACTION广播事件的响应,因此,在onPause()方法中注销该广播接收者(myBroadcastReceiver)。

请确保在发送广播之前已经注册了相应的广播接收者,否则广播接收者将无法接收到广播。可以通过静态注册(在AndroidManifest.xml文件中声明广播接收者)或动态注册(在代码中调用registerReceiver()方法)来注册广播接收者。

当广播发送后,系统将会将广播传递给所有匹配的接收者,并调用它们的onReceive()方法来处理广播事件。接收者可以通过获取Intent对象中的数据来执行相应的操作。

MyBroadcastReceiverApp项目对应的布局如【文件8-4】所示。

【文件8-4】 activity_main.xml

```xml
<?xml version = "1.0" encoding = "utf-8"?>
< RelativeLayout xmlns:android = "http://schemas.android.com/apk/res/android"
```

```
        xmlns:app = "http://schemas.android.com/apk/res - auto"
        xmlns:tools = "http://schemas.android.com/tools"
        android:layout_width = "match_parent"
        android:layout_height = "match_parent"
        tools:context = ".MainActivity">
        < Button
            android:id = "@ + id/btn_selfdefinebroadcast"
            android:layout_width = "wrap_content"
            android:layout_height = "wrap_content"
            android:text = "点击发送自定义广播"
            android:layout_centerInParent = "true"/>
</RelativeLayout >
```

activity_main.xml 采用相对布局,在其中定义一个 Button,id 为 btn_selfdefinebroadcast,text 文本为"点击发送自定义广播",居中显示。

最终,程序运行结果如图 8.3、图 8.4 所示。

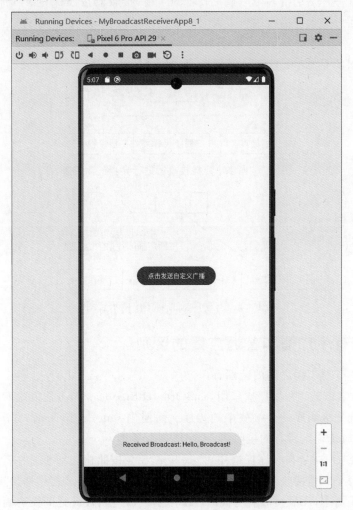

图 8.3　运行结果

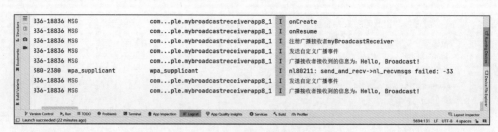

图 8.4　运行过程中 LogCat 的输出

观看视频

8.3　有序广播

有序广播(Ordered Broadcast)和普通广播(Normal Broadcast)是 Android 中两种不同的广播机制,它们在广播的发送和接收方式上有所不同。8.2 节所列示例程序为普通广播。

有序广播是一种发送给多个广播接收者的广播,接收器按照优先级顺序一个接一个地接收和处理广播。每个广播接收者可以对广播进行拦截、修改、终止或将广播传递给下一个接收器。

图 8.5 和图 8.6 显示异步正常广播与有序广播顺序传播的异同。

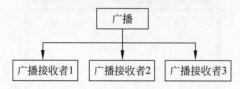

图 8.5　向多个广播接收者进行异步正常广播

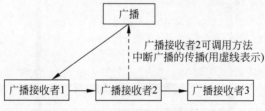

图 8.6　有序广播的顺序传播示意图

8.3.1　有序广播和普通广播的区别

有序广播和普通广播之间的区别有:

(1) 广播发送方式:有序广播采用 sendOrderedBroadcast()方法发送有序广播,广播会按照接收器的优先级顺序一个接一个地传递。而使用 sendBroadcast()方法发送的为普通广播,广播会同时传递给所有的接收器,无序地进行处理。

(2) 广播接收顺序:在有序广播中,广播接收者会按照优先级顺序一个接一个地接收和处理广播。每个广播接收者可以对广播进行拦截、修改、终止或将广播传递给下一个接收者。而在普通广播中,广播接收者为同时、无序地接收广播,彼此之间没有固定的顺序。

(3) 广播的传递:在有序广播中系统会按照接收器的优先级顺序依次调用每个接收器的 onReceive()方法,传递广播和数据。如果某个接收者调用了 abortBroadcast()方法,则后续的接收器将无法接收到该广播。在普通广播中,系统会同时将广播传递给所有的接收器,彼此之间没有影响。

选择使用有序广播还是普通广播取决于实际应用场景的需求。有序广播适用于需要控制广播传递顺序、进行广播拦截、修改和终止的场景。而普通广播则适用于不需要特定顺序和处理逻辑的场景,广播接收器之间相互独立。

8.3.2 有序广播的发送与处理流程

有序广播的发送和处理过程如下:

(1) 发送有序广播:使用 sendOrderedBroadcast()方法发送有序广播。该方法接受一个 Intent 对象作为参数,并可选择指定接收器权限、结果接收器和初始数据。

```
Intent intent = new Intent("com.example.chapter8.MY_CUSTOM_ACTION");
sendOrderedBroadcast(intent, null);
```

(2) 接收有序广播:创建广播接收器类并注册接收器。接收器的优先级通过 android:priority 属性在静态注册时设置或 setPriority()方法在动态注册时设置。在接收器的 onReceive()方法中,可以处理接收到的广播事件;可以检查广播动作、获取额外数据,并根据需要执行相应操作;还可以使用 abortBroadcast()方法终止广播的传递,使后续接收器无法接收到该广播。

```
public class MyBroadcastReceiver extends BroadcastReceiver {
    @Override
    public void onReceive(Context context, Intent intent) {
        //处理接收到的广播事件
        if (intent.getAction().equals("com.example.chapter8.MY_CUSTOM_ACTION")) {
            //处理自定义广播事件
            String message = intent.getStringExtra("message");
            //执行相应的操作
        }
        //终止广播
        abortBroadcast();
    }
}
```

开发人员要注意,Android 系统按照接收器的优先级顺序依次调用每个广播接收者的 onReceive()方法,传递广播和数据。如果某个接收者调用了 abortBroadcast()方法,则后续的接收器将无法接收到该广播。

有序广播的优点是可以控制广播的传递顺序,并且接收器可以对广播进行拦截、修改和终止。但要注意,滥用有序广播可能会导致性能问题和广播链的复杂性,因此应谨慎使用,并确保合理处理广播的传递和处理逻辑。

8.3.3 有序广播实例

在 MyOrderedBroadcastReceiverApp 项目实例中,定义了三个广播接收者,分别为 MyOrderedBroadcastReceiver1、MyOrderedBroadcastReceiver2、MyOrderedBroadcastReceiver3,设定 priority 优先级分别为 0、100、50。之后在 MainActivity 中定义一个 Button,并设置监听器为

MainActivity(实现 View.OnClickListener 接口),重写了父接口 View.OnClickListener 中的 onClick()方法。在 onClick()方法中发送有序广播,三个广播接收者接收广播并做出处理。

MyOrderedBroadcastReceiver1 代码如【文件 8-5】所示。

【文件 8-5】 MyOrderedBroadcastReceiver1.java

```
package com.example.myorderedbroadcastreceiverapp;
import android.content.BroadcastReceiver;
import android.content.Context;
import android.content.Intent;
import android.util.Log;
import android.widget.Toast;
public class MyOrderedBroadcastReceiver1 extends BroadcastReceiver {
    @Override
    public void onReceive(Context context, Intent intent) {
        Log.i("MSG","MyOrderedBroadcastReceiver1.onReceive");
        Log.i("MSG","intent.getAction()" + intent.getAction());
        //在接收到广播时执行的操作
        if (intent.getAction().equals("com.example.chapter08.MY_CUSTOM_ACTION")){
            //处理自定义广播事件
            String message = intent.getStringExtra("message");
            Log.i("MSG","广播接收者接收到的信息为:" + message);
            Toast.makeText(context, "Received Broadcast: " + message,Toast.LENGTH_SHORT).show();
        }
        //终止广播
        //abortBroadcast();
    }
}
```

MyOrderedBroadcastReceiver2、MyOrderedBroadcastReceiver3 与接收者 1 的代码相同,这里不再赘述。

在 MainActivity 中编写对应代码,设置优先级分别为 0、100、50(设置自定义广播接收者 1、自定义广播接收者 2、自定义广播接收者 3 分别为 0、50、100,方便观察广播接收到的顺序),代码如【文件 8-6】所示。

【文件 8-6】 MainActivity.java

```
package com.example.myorderedbroadcastreceiverapp;
……//省略导入的相关包、类
public class MainActivity extends AppCompatActivity implements View.OnClickListener{
    private Button mButton;
    private MyOrderedBroadcastReceiver1 myOrderedBroadcastReceiver1;
    private MyOrderedBroadcastReceiver2 myOrderedBroadcastReceiver2;
    private MyOrderedBroadcastReceiver3 myOrderedBroadcastReceiver3;
    @Override
    protected void onCreate(Bundle savedInstanceState) {
        super.onCreate(savedInstanceState);
        Log.i("MSG","onCreate");
        setContentView(R.layout.activity_main);
        mButton = findViewById(R.id.btn_orderedbroadcast);
        mButton.setOnClickListener(this);
```

```java
        myOrderedBroadcastReceiver1 = new MyOrderedBroadcastReceiver1();
        myOrderedBroadcastReceiver2 = new MyOrderedBroadcastReceiver2();
        myOrderedBroadcastReceiver3 = new MyOrderedBroadcastReceiver3();
    }

    @Override
    protected void onResume() {
        super.onResume();
        Log.i("MSG","onResume");
        registerReceivers(myOrderedBroadcastReceiver1,0);
        registerReceivers(myOrderedBroadcastReceiver2,100);
        registerReceivers(myOrderedBroadcastReceiver3,50);
    }
    //为了方便调用
    private void registerReceivers(BroadcastReceiver receiver, int priority){
        IntentFilter intentFilter = new IntentFilter();
        intentFilter.setPriority(priority);
        intentFilter.addAction("com.example.chapter08.MY_CUSTOM_ACTION");
        registerReceiver(receiver, intentFilter);
        Log.i("MSG","注册广播接收者:" + receiver);
    }


    private void unregisterReceivers(){
        unregisterReceiver(myOrderedBroadcastReceiver1);
        unregisterReceiver(myOrderedBroadcastReceiver2);
        unregisterReceiver(myOrderedBroadcastReceiver3);
    }

    @Override
    protected void onPause() {
        super.onPause();
        Log.i("MSG","onPause");
        unregisterReceivers();
        Log.i("MSG","注销广播接收者 myBroadcastReceiver");
    }

    @Override
    public void onClick(View v) {
        Intent intent = new Intent();
        intent.setAction("com.example.chapter08.MY_CUSTOM_ACTION");
        //添加额外的数据到 Intent 对象
        intent.putExtra("message", "Hello, OrderedBroadcast!");
        //发送广播
        sendOrderedBroadcast(intent,null);
        Log.i("MSG","发送有序广播");
    }
}
```

　　MainActivity.java 的 onCreate()方法中定义了三个广播接收者，在 onResume()方法中调用自定义的 registerReceivers()方法。registerReceivers()方法中第二个参数为对应广播接收者的优先级。在这里传递的三个广播接收者的 priority 优先级参数分别为 0、100、

50，数值越大，越优先监听到广播。因此，广播接收顺序为 2、3、1。

通过查阅 Android 开发文档，可以看出，priority 优先级的取值默认值为 0，取值范围为 [-1000,1000]。开发文档中分别定义为静态整型常量 SYSTEM_LOW_PRIORITY 和 SYSTEM_HIGH_PRIORITY。

同时，在 MainActivity 中重写了 View.onClickListener 接口的 onClick()方法，调用 sendOrderedBroadcast()方法发送有序广播。

三个广播接收者接收到广播之后，在 LogCat 下做相应的输出。在不调用 abortBroadcast()方法时，输出顺序为 MyOrderedBroadcastReceiver2、MyOrderedBroadcastReceiver3、MyOrderedBroadcastReceiver1，如图 8.7 所示。

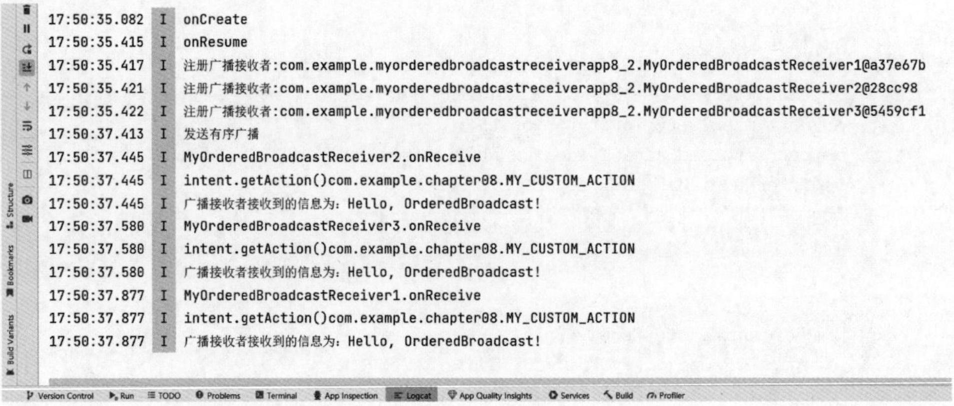

图 8.7　LogCat 窗口 1

如果 abortBroadcast()方法没有被注释掉，就会仅有 MyOrderedBroadcastReceiver2 接收到广播，而 MyOrderedBroadcastReceiver3、MyOrderedBroadcastReceiver1 由于广播被中止而收不到广播，运行结果如图 8.8 所示。

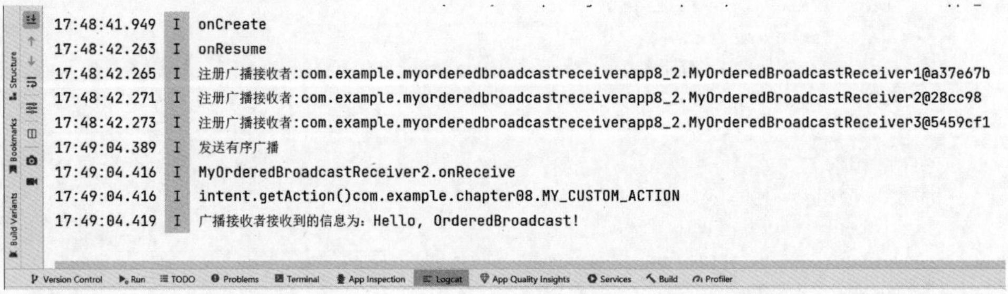

图 8.8　LogCat 窗口 2

8.4　系统预定义广播

Android 提供了许多预定义的广播类型，可以用于接收和处理各种系统级别的事件和状态变化。

以下列出一些常见的 Android 预设广播类型。

（1）系统广播（System Broadcasts）。

- ACTION_BOOT_COMPLETED：设备启动完成时发送的广播。
- ACTION_POWER_CONNECTED：设备充电器连接时发送的广播。
- ACTION_POWER_DISCONNECTED：设备充电器断开连接时发送的广播。
- ACTION_SHUTDOWN：设备关机时发送的广播。

(2) 网络状态广播(Network State Broadcasts)。
- ACTION_CONNECTIVITY_CHANGE：网络连接状态发生变化时发送的广播。
- ACTION_Wi-Fi_STATE_CHANGED：Wi-Fi 状态发生变化时发送的广播。
- ACTION_NETWORK_STATE_CHANGED：网络状态发生变化时发送的广播。

(3) 电池状态广播(Battery State Broadcasts)。
- ACTION_BATTERY_CHANGED：电池状态发生变化时发送的广播。
- ACTION_BATTERY_LOW：电池电量过低时发送的广播。
- ACTION_BATTERY_OKAY：电池电量恢复正常时发送的广播。

(4) 时间广播(Time Broadcasts)。
- ACTION_TIME_TICK：每分钟发送一次的广播。
- ACTION_DATE_CHANGED：日期发生变化时发送的广播。
- ACTION_TIMEZONE_CHANGED：时区发生变化时发送的广播。

(5) 屏幕状态广播(Screen State Broadcasts)。
- ACTION_SCREEN_ON：屏幕开启时发送的广播。
- ACTION_SCREEN_OFF：屏幕关闭时发送的广播。
- ACTION_USER_PRESENT：用户解锁屏幕时发送的广播。

(6) 媒体广播(Media Broadcasts)。
- ACTION_MEDIA_MOUNTED：外部存储设备挂载时发送的广播。
- ACTION_MEDIA_UNMOUNTED：外部存储设备卸载时发送的广播。
- ACTION_MEDIA_SCANNER_STARTED：开始扫描媒体文件时发送的广播。
- ACTION_MEDIA_SCANNER_FINISHED：媒体文件扫描完成时发送的广播。

这里列出部分 Android 系统中预设的广播类型，还有其他许多广播类型可用于接收和处理不同的系统事件和状态变化。Android 预设的广播源代码可以在 Android 开源项目中找到。在源代码中包括了各种系统组件和功能的实现。

读者可以通过查阅 Android 开发文档，了解更多关于广播类型和相关信息的详细信息。

8.5 应用实例：通过广播机制判断手机电量状态

观看视频

要判断 Android 手机的电量状态，可以通过监听系统发送的电量变化广播来实现。以下示例代码实现了用于接收并处理电量变化的广播 App。

(1) 创建一个名为 BatteryStatusApp 的应用程序，指定包名为 com.example.batterystatusapp，之后设计用户交互界面。程序主界面对应的布局具体代码详见【文件 8-7】。

【文件 8-7】 activity_main.xml

```xml
<?xml version = "1.0" encoding = "utf-8"?>
<androidx.constraintlayout.widget.ConstraintLayout xmlns:android = "http://schemas.android.com/apk/res/android"
    xmlns:app = "http://schemas.android.com/apk/res-auto"
    xmlns:tools = "http://schemas.android.com/tools"
    android:layout_width = "match_parent"
    android:layout_height = "match_parent"
    tools:context = ".MainActivity">

    <TextView
        android:id = "@+id/tv_battery"
        android:layout_width = "wrap_content"
        android:layout_height =."wrap_content"
        android:text = "测试电量变化广播!"
        android:textColor = "@color/black"
        android:textSize = "36sp"
        app:layout_constraintBottom_toBottomOf = "parent"
        app:layout_constraintEnd_toEndOf = "parent"
        app:layout_constraintStart_toStartOf = "parent"
        app:layout_constraintTop_toTopOf = "parent" />

</androidx.constraintlayout.widget.ConstraintLayout>
```

在 activity_main.xml 文件中,设置布局为默认约束布局,放置了一个控件 TextView,提示本程序为测试接收到电量变化的广播。

(2) 创建广播接收者类(BatteryStatusReceiver),这个广播接收者可以接收到电池电量发生变化的广播,并使用 Toast 和 LogCat 的方式输出提示信息。在 batterystatusapp 中定义了广播接收者 BatteryStatusReceiver 和主活动 MainActivity,接收手机电池状态是否改变。所以,这里监听的是 Intent 中的静态常量字符串 ACTION_BATTERY_CHANGED。之后在 BatteryStatusReceiver 的 onReceive()方法中做电池电量比例、是否处于充电状态等信息的输出。代码详见【文件 8-8】。

【文件 8-8】 BatteryStatusReceiver.java

```java
package com.example.batterystatusapp;
import android.content.BroadcastReceiver;
import android.content.Context;
import android.content.Intent;
import android.os.BatteryManager;
import android.util.Log;
import android.widget.Toast;
public class BatteryStatusReceiver extends BroadcastReceiver {
    @Override
    public void onReceive(Context context, Intent intent) {
        if (intent.getAction().equals(Intent.ACTION_BATTERY_CHANGED)) {
            int level = intent.getIntExtra(BatteryManager.EXTRA_LEVEL, -1);
            int scale = intent.getIntExtra(BatteryManager.EXTRA_SCALE, -1);
            int batteryLevel = (int) ((level / (float) scale) * 100);

            int status = intent.getIntExtra(BatteryManager.EXTRA_STATUS, -1);
```

```
            boolean isCharging = status ==
        BatteryManager.BATTERY_STATUS_CHARGING||
        status == BatteryManager.BATTERY_STATUS_FULL;

            if (isCharging) {
                //正在充电
                Toast.makeText(context, "Battery is charging",
                            Toast.LENGTH_SHORT).show();
            } else {
                //未在充电
                Toast.makeText(context, "Battery is not charging",
                            Toast.LENGTH_SHORT).show();
            }

            //打印电量信息
            Log.d("MSG", "Battery Level: " + batteryLevel + "%");
        }
    }
}
```

在上述代码中，创建了一个继承自 BroadcastReceiver 类的广播接收者 BatteryStatusReceiver。在 onReceive()方法中，检查接收到的广播是否为电池状态发生变化的广播(Intent.ACTION_BATTERY_CHANGED)。如果是，则可以通过 getIntExtra()方法获取电量相关的信息，如电量水平(BatteryManager.EXTRA_LEVEL)和电量最大值(BatteryManager.EXTRA_SCALE)。通过计算得到电量百分比，然后获取充电状态(BatteryManager.EXTRA_STATUS)并判断手机是否正在充电。根据充电状态，可以执行相应的操作，如显示 Toast 提示。

(3) 注册广播接收者(在 MainActivity.java 中完成)。

在 Intent.java 中明确告知开发者监听电池类广播在 Android 系统中是不支持使用静态注册的，只能使用动态注册的方法才可以监听到对应的广播，文档如图 8.9 所示。

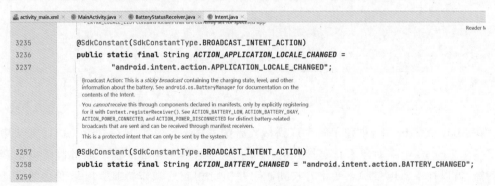

图 8.9　监听电池类广播需要在动态注册源代码中说明

【文件 8-9】是 MainActivity 程序源代码，其中读者应重点关注如何进行广播接收器的动态注册。

【文件 8-9】　MainActivity.java

```java
package com.example.batterystatusapp;
import androidx.appcompat.app.AppCompatActivity;
import android.content.Intent;
import android.content.IntentFilter;
import android.os.Bundle;
import android.util.Log;
import android.view.View;
import android.widget.Toast;
public class MainActivity extends AppCompatActivity {
    BatteryStatusReceiver receiver;
    @Override
    protected void onCreate(Bundle savedInstanceState) {
        super.onCreate(savedInstanceState);
        setContentView(R.layout.activity_main);
        findViewById(R.id.tv_battery).setOnClickListener(new View.OnClickListener(){
            @Override
            public void onClick(View v) {
                Toast.makeText(MainActivity.this, "显示广播侦测当前电量",
                        Toast.LENGTH_SHORT).show();
            }
        });
    }

    @Override
    protected void onResume() {
        super.onResume();
        Log.i("MSG","onResume");
        //创建广播接收器实例
        receiver = new BatteryStatusReceiver();
        //创建 IntentFilter 对象,指定接收的广播动作
        IntentFilter filter = new IntentFilter(Intent.ACTION_BATTERY_CHANGED);
        //注册广播接收器
        registerReceiver(receiver, filter);
    }

    @Override
    protected void onPause() {
        super.onPause();
        Log.i("MSG","onPause");
        unregisterReceiver(receiver);
    }
}
```

在 MainActivity 中,创建了一个广播接收者的实例 receiver,并在 onResult()方法中创建了一个 IntentFilter 对象来指定接收的广播动作为电量变化的广播。通过监听电量变化的广播,可以在手机电量状态变化时收到通知并执行相应的操作。监听电量变化的广播在 Android 中定义为常量字符串 Intent.ACTION_BATTERY_CHANGED,之后调用 registerReceiver()方法来动态注册广播接收器,以便它可以接收到电量变化的广播。

```java
public static final String ACTION_BATTERY_CHANGED =
"android.intent.action.BATTERY_CHANGED";
```

这里要注意,在不需要接收广播时,要记得调用 unregisterReceiver()方法来注销广播接收器,以避免内存泄漏。

在图 8.10～图 8.11 的手机扩展控制面板中,读者通过设置 Charge level 为不同的百分比,可以在 LogCat 中看到电量的变化。可以看出,Battery status 为 Unknown/Charging/Discharging/Not charging/Full 不同值时,接收到的电池状态在"Battery is charging"和"Battery is not charging"之间切换。

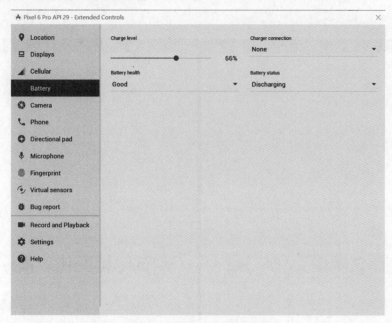

图 8.10　手机扩展控制面板

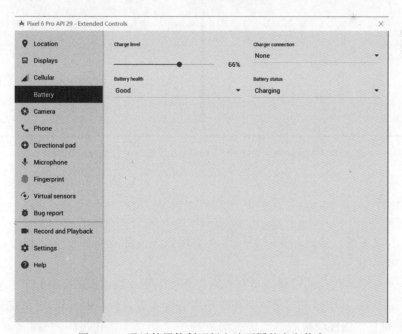

图 8.11　通过扩展控制面板电池不同的充电状态

图 8.12 为 LogCat 下输出的电池电量,输出当前电量为 66%。

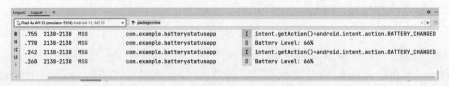

图 8.12　LogCat 下输出的电量信息

图 8.13 为是否充电时的 Toast 输出。

图 8.13　MainActivity 下使用 Toast 方式输出电量变化广播信息

8.6　小结

- Android 系统广播机制是一种用于在应用程序内部或与其他应用程序之间进行通信的重要机制。
- Android 系统包括的广播类型有标准广播(Normal Broadcast)和有序广播(Ordered Broadcast),其中标准广播同时传递给所有接收者,接收者之间没有固定顺序。有序广播则是广播按照接收者的优先级顺序一个接一个地传递,每个接收器可以拦截、修改、终止广播或将广播传递给下一个接收器。

- 使用 sendBroadcast()方法发送标准广播,使用 sendOrderedBroadcast()方法发送有序广播。
- 广播注册的方法分为静态注册和动态注册,静态注册是在 AndroidManifest.xml 文件中使用 receiver 元素进行静态注册,而动态注册是在代码中使用 registerReceiver()方法编程注册。
- 广播接收者(BroadcastReceiver)需要继承自 BroadcastReceiver 类,并通过重写父类的 onReceive()方法来处理接收到的广播,同时根据软件需求的需要,在接收到广播时执行特定的操作,如更新 UI、启动服务、发送通知等。
- 广播过滤器(IntentFilter)用于指定广播接收器接收的广播类型,可以通过指定广播动作(Action)、数据类型(Data)和类别(Category)等来定义过滤条件。
- Android 系统中提供了许多预定义的广播类型,如系统广播、网络状态广播、电池状态广播等。
- 通过使用广播机制,Android 应用程序可以实现与系统和其他应用程序之间的通信,从而实现各种功能和交互。开发人员可以根据自己的需求选择适当的广播类型,并编写广播接收器来处理接收到的广播。

习题 8

1. 请查阅资料,编写代码演示观察者模式。
2. 结合自身常用软件,列出软件中可能使用到的预定义广播。
3. 编写广播接收者,对 Wi-Fi 是否连接或断开进行判定,并输出提示信息。

第9章 Service

本章导图

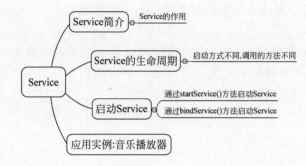

主要内容

- Service 的基本概念。
- Service 的应用场景。
- 两种启动 Service 方式的区别。

难点

- Service 的应用场景。
- Service 生命周期中方法的回调。

Service 是 Android 系统中的四大组件（Activity、Service、BroadcastReceiver、ContentProvider）之一，平时也称作"服务"。在开发人员的角度称"服务"或者 Service 的意义是相同的，在本书中统一称为 Service。它不能自己运行只能后台运行，并且可以和其他组件进行交互。任何 Android 应用程序中需要长时间或有阻塞操作的部分都可以在 Service 中运行，这样主线程就可以尽可能地保持独立。虽然开发者可以在 Activity 中启动新的线程进行部分后台操作，但官方推荐的方法是将后台操作移到单独的 Service 中。Android 中的 Service 组件提供了一种有效的方法，将后台任务的应用逻辑与处理用户界面的代码分离开来。

本章介绍 Android 系统中有关 Service 的基本概念，描述 Service 的生命周期、Service 的应用场景、Service 的注册（清单文件注册、代码注册）、Service 的启动以及如何与其他组件进行通信等。

9.1 Service 简介

在 Android 的官方文档（https://developer.android.com/reference/android/app/Service）中，对 Service 组件做了如下描述：Service 是一个应用程序组件，表示应用程序希望在不与用户交互的情况下执行长时间运行的操作，或者提供功能供其他应用程序使用。

每个 Service 类要求在 AndroidManifest.xml 中必须有一个相应的 <service> 声明。Service 可以通过 Context.startService() 或 Context.bindService() 启动。和其他应用程序对象一样，Service 在其托管进程的主线程中运行。这意味着，如果 Service 要做任何 CPU 密集型（如 MP3 播放）或阻塞（如联网）操作，它应该生成自己的线程来完成这些工作。

Service 不是独立的进程，Service 对象本身并不意味着它在自己的进程中运行，它所属的应用程序在同一进程中运行。

Service 不是线程，它本身并不是在主线程之外完成工作的一种手段（以避免应用程序不响应错误）。

Service 本身实际提供了如下两个主要特性：

(1) 应用程序告诉 Android 系统它想在后台做的事情（即使用户不直接与应用程序交互）。这对应于对 Context.startService() 方法的调用，它要求系统为 Service 调度工作，直到 Service 或其他人显式停止它。

(2) 应用程序向其他应用程序公开其部分功能的工具。这对应于 Context.bindService() 的调用，该调用允许与 Service 建立长期连接，以便与 Service 进行交互。

当一个 Service 组件被实际创建时，系统实际上所做的只是实例化该组件，并在主线程上调用 Service 的 onCreate() 方法和任何其他适当的回调。Service 要用适当的行为来实现这些功能，例如创建一个辅助线程来完成它的工作。

9.2 Service 的生命周期

Android 中 Service 具有自己的生命周期，它通过一系列的生命周期方法来管理其状态和执行操作，具体如图 9.1 所示。在 Service 的生命周期中，随着启动状态的不同，会调用不同的方法。如使用 startService() 启动 Service，依次调用 onCreate()、onStart()、onDestroy() 三个方法。如使用 bindService() 启动 Service，则依次调用的为 onCreate()、onBind()、onUnbind()、onDestroy() 方法。

这里需要注意的是，如果 Service 已经启动了，当再次启动 Service 时，不会再执行 onCreate() 方法，而是直接执行 onStart() 或者 onBind() 方法。它可以通过 Service.stopSelf() 方法或者 Service.stopSelfResult() 方法来停止自己，只要调用一次 stopService() 方法便可以停止 Service。图 9.1 是 Service 的生命周期。

(1) 创建（onCreate）：当 Service 被创建时，系统会调用 Service 的 onCreate() 方法。在这个方法中，可以进行初始化操作，例如初始化变量、设置资源等。这个方法只会在 Service 创建时调用一次。

(2) 启动（onStartCommand）：通过调用 startService() 方法启动 Service 时，系统会调

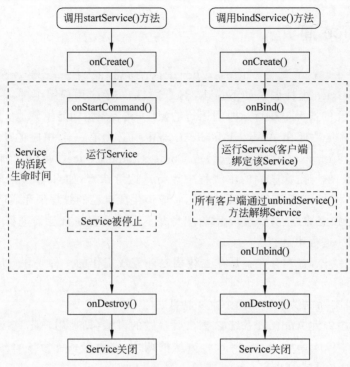

图 9.1 Service 的生命周期

用 onStartCommand()方法。在这个方法中,可以执行 Service 的实际操作,如执行网络请求、下载文件等。该方法执行完毕需要返回一个整数值,该值表示系统在停止 Service 后应该采取的行动。这个整数值被称为"启动标志"(Start Flag)。

以下是 onStartCommand()方法的返回值的意义。

- START_STICKY_COMPATIBILITY(代表数值 0):在旧版 Android 中使用的标志,与 START_STICKY 类似,但是在低内存情况下不会保证 Service 被重新创建。
- START_STICKY(代表数值 1):如果系统在执行完 Service 的 onStartCommand()方法后被意外终止,系统会重新创建 Service,并调用 onStartCommand()方法,但不会重新传递之前的 Intent。这对于执行周期性任务很有用,例如定时检查更新。
- START_NOT_STICKY(代表数值 2):如果系统在执行完 Service 的 onStartCommand()方法后被意外终止,系统不会重新创建 Service,直到再次调用 startService()方法。这对于执行一次性任务很有用,例如下载文件。
- START_REDELIVER_INTENT(代表数值 3):如果系统在执行完 Service 的 onStartCommand()方法后被意外终止,系统会重新创建 Service,并重新传递之前的 Intent 给 onStartCommand()方法。这对于需要确保任务完成的情况很有用,例如上传文件。

程序开发人员需要根据的 Service 的性质和需求选择适当的返回值。如果 Service 执行周期性任务,可能会选择 START_STICKY。如果 Service 执行一次性任务,可能会选择 START_NOT_STICKY。如果 Service 需要确保任务完成,可能会选择 START_REDELIVER_INTENT。

（3）绑定（onBind）：如果希望其他组件可以绑定到 Service，需要在 Service 中实现 onBind()方法。通过 bindService()方法可以让其他组件与 Service 建立连接，这种连接是一个客户端/服务器模式的连接，使得客户端可以调用 Service 的方法。

（4）运行（运行时）：一旦 Service 被启动，它将一直保持运行状态，直到调用 stopService()或者 stopSelf()方法来停止 Service。在这期间，Service 会执行 onStartCommand()方法中的操作。

（5）销毁（onDestroy）：当 Service 不再需要运行，或者通过调用 stopService()或者 stopSelf()停止 Service 时，系统会调用 onDestroy()方法。在这个方法中，可以进行资源的释放和清理操作，例如关闭文件、释放内存等。这是 Service 的最后一个生命周期方法。

需要注意的是，Service 在默认情况下运行在应用的主线程中，因此如果 Service 中执行耗时操作，可能会导致应用的主线程阻塞，影响应用的响应性。为了避免这种情况，可以在 Service 中使用线程、异步任务，或者更现代的方法如 Kotlin 协程等来处理耗时操作。

此外，前台 Service（Foreground Service）是一种特殊的 Service，需要在状态栏显示通知，以保证系统不会轻易地终止该 Service，从而提供更好的用户体验。在前台 Service 中，需要在适当时创建和移除通知，并确保 Service 处于可见状态。

9.3 启动 Service

可以使用 startService()和 bindService()两种不同的方式来启动或与 Service 建立连接。它们在 Service 的生命周期、通信方式以及用途上有一些不同。

1. 启动方式

（1）startService()：使用 startService()方式启动 Service，Service 会被启动并开始执行它的工作。即使调用者（通常是 Activity）退出，Service 仍然会继续运行，除非调用 stopService()方法或者 Service 内部调用了 stopSelf()方法来停止自己。

（2）bindService()：使用 bindService()方式启动 Service，Service 会绑定到调用者，并且只有调用者与 Service 处于绑定状态时，Service 才会一直运行。当所有绑定的客户端（通常是 Activity）断开连接时，Service 会被销毁。

2. 生命周期

startService()：Service 会执行 onCreate()和 onStartCommand()方法，然后进入运行状态。它会一直运行，直到调用 stopService()方法或者 Service 内部调用了 stopSelf()方法。

bindService()：Service 的生命周期与绑定的客户端相关联。当所有绑定的客户端断开连接时，系统会调用 Service 的 onUnbind()方法，然后根据情况调用 onRebind()方法或者销毁 Service。

3. 通信方式

（1）startService()：通常用于启动执行一次性任务的 Service。调用者可以通过 Intent 将参数传递给服务，但与 Service 的通信相对简单。

（2）bindService()：用于建立客户端与 Service 之间的连接，使得调用者可以获取 Service 的实例，从而调用 Service 的方法。它更适合于复杂的、持续性的交互。

4. 停止 Service

（1）startService()：调用 stopService()方法或者在 Service 内部调用 stopSelf()方法来停止 Service。

（2）bindService()：调用 unbindService()方法解除绑定，并在所有绑定的客户端都解除连接时，Service 会被销毁。

注意，stopSelf()方法是在 Service 内部调用的。它用于在 Service 内部主动停止自身的运行。这个方法通常在 Service 完成了它的任务后调用，或者在 Service 遇到需要停止的条件时调用。调用 stopSelf()方法会触发系统调用 Service 的 onDestroy()方法来执行清理操作。因此，可以在 onDestroy()方法中释放资源、取消定时任务等。stopSelf()方法是在 Service 内部调用的，用于在 Service 内部主动停止自身。调用 stopSelf()方法会导致 Service 的 onDestroy()方法被调用，从而执行清理操作。而 onDestroy()方法则是生命周期方法，在 Service 即将被销毁时被系统调用，用于执行最后的清理操作。

总的来说，如果需要在后台执行一次性任务，或者需要 Service 在调用者退出后继续运行，则可以使用 startService()方法。如果需要与 Service 进行复杂的交互，获取 Service 实例，或者需要在调用者与 Service 之间建立连接，则可以使用 bindService()方法。

9.3.1 创建、配置 Service

Service 的创建方式同 Activity 及其他组件的操作一样，首先在 Android Studio 中创建一个工程，这里命名为 ServiceApp9_1，包名为 com.example.serviceapp，接着在程序的包名处（应用程序已经存在）右击，选择 New→Service→Service 选项，如图 9.2 所示。

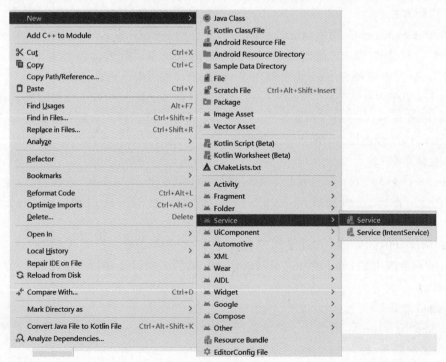

图 9.2 创建 Service

在下一步中,确定类名为 MyService、Exported 和 Enabled 选项已经勾选、语言为 Java 后,单击 Finish 按钮完成 Service 的创建,如图 9.3 所示。

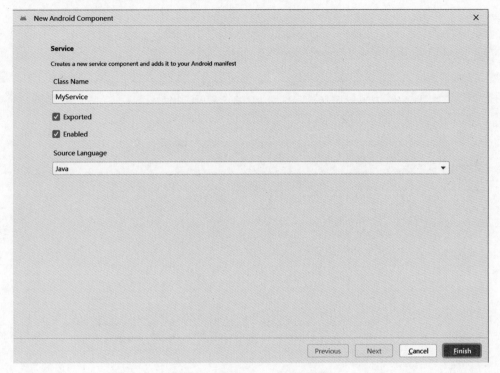

图 9.3 确认 Service 相关信息

根据图 9.2 和图 9.3 中的步骤创建完成的 MyService 的具体代码如【文件 9-1】所示。

【文件 9-1】 MyService.java

```
package com.example.serviceapp
import android.app.Service;
import android.content.Intent;
import android.os.IBinder;
public class MyService extends Service {
    public MyService() {
    }
    @Override
    public IBinder onBind(Intent intent) {
        throw new UnsupportedOperationException("Not yet implemented");
    }
}
```

在上述代码中可以看到,MyService 派生自 Service 类,实现了一个无参构造方法 MyService()。跟踪源代码可以知道 Service 是一个抽象类(Abstract Class),其中定义 onBind()为抽象方法,因此在 Service 的子类(MyService)中,必须对该方法进行 Override (即覆盖父类的方法并对其重写)。因为目前为代码自动生成状态,所以该方法目前如果调用,则会抛出一个不支持的操作异常。

同时,Android Studio 也会自动在 AndriodManifest.xml 文件中对 MyService 组件进

行注册,代码如【文件 9-2】所示。

【文件 9-2】 AndriodManifest.xml

```xml
<?xml version = "1.0" encoding = "utf-8"?>
<manifest xmlns:android = "http://schemas.android.com/apk/res/android"
    xmlns:tools = "http://schemas.android.com/tools">
    <application android:allowBackup = "true"
        android:dataExtractionRules = "@xml/data_extraction_rules"
        android:fullBackupContent = "@xml/backup_rules"
        android:icon = "@mipmap/ic_launcher"
        android:label = "@string/app_name"
        android:roundIcon = "@mipmap/ic_launcher_round"
        android:supportsRtl = "true"
        android:theme = "@style/Theme.ServiceApp"
        tools:targetApi = "31">
        <service
            android:name = ".MyService"
            android:enabled = "true"
            android:exported = "true"></service>
        <activity
            android:name = ".MainActivity"
            android:exported = "true">
            <intent-filter>
                <action android:name = "android.intent.action.MAIN" />
                <category android:name = "android.intent.category.LAUNCHER" />
            </intent-filter>
        </activity>
    </application>
</manifest>
```

在上述 AndroidManifest.xml 文件中,service 标签用于声明应用中的 Service 组件,它定义了 Service 的属性和特性。这里设置了 android:name、android:enabled、android:exported 三个属性。其中,android:name 指定 Service 类的完整名称,包括包名,这里省略了包名,仅写了类名;android:enabled 指定 Service 是否可用,这里设置为 true,表示 Service 可用;android:exported 指定 Service 是否可以被其他应用程序的组件访问,若设置为 true,则其他应用程序可以通过显式或隐式 Intent 访问该 Service,这里设置为 true。

除此之外,service 标签中还有一些其他参数。

(1) android:process:指定 Service 运行的进程名称。如果不指定,则 Service 将在默认的主进程中运行。

(2) android:permission:指定访问 Service 的权限名称。只有具有指定权限的应用程序才能访问该 Service。

(3) android:isolatedProcess:指定 Service 是否在自己的进程中运行。如果设置为 true,则 Service 将在一个与主应用进程隔离的进程中运行。

(4) android:description:提供对 Service 的简要描述,用于解释 Service 的作用或功能。

(5) android:label:指定一个字符串资源作为 Service 的标签,用于显示在应用程序的用户界面中。

(6) android:icon:指定一个图标资源,用于表示 Service。

（7）android：priority：指定 Service 的优先级，用于在有限资源情况下决定哪个 Service 会更容易被保留。

（8）android：foregroundServiceType：用于设置前台 Service 的类型。可以是 dataSync、location、mediaPlayback 等类型，以便系统知道如何优化这种前台 Service 的行为。

这些参数允许在清单文件中对 Service 进行详细配置，定义 Service 的可见性、运行方式以及与其他应用程序的交互方式。根据应用需求，可以根据实际情况设置这些参数。

9.3.2 启动和停止 Service

观看视频

系统可以通过两种方式启动 Service：一种是调用 Context.startService()方法；另一种是调用 Context.bindService()方法。如果使用 Context.startService()方法启动 Service，那么系统将检索 Service（创建对应的 Service 并在需要时调用它的 onCreate()方法），然后使用客户端提供的参数调用它的 onStartCommand(Intent，int，int)方法。

对于已启动的 Service，它们可以决定运行另外两种主要的操作模式，这取决于它们从 onStartCommand()方法返回的值：START_STICKY 用于根据需要显式启动和停止的 Service；START_NOT_STICKY 或 START_REDELIVER_INTENT 用于只在处理发送给它们的任何命令时应该保持运行的 Service。有关语义的更多详细信息，请参阅相关文档。

客户端也可以使用 Context.bindService()方法来获取到 Service 的持久连接。如果 Service 还没有运行（在运行时调用 onCreate()方法），这同样会创建 Service，但不会调用 onStartCommand()方法。客户端将接收 Service 从其 onBind(Intent)方法返回的 IBinder 对象，允许客户端随后回调 Service。只要连接建立（无论客户端是否在服务的 IBinder 上保留引用），Service 将保持运行。通常，返回的 IBinder 用于用 AIDL 编写的复杂接口。

Service 可以被启动，也可以被连接绑定。在这种情况下，只要 Service 启动，或者有一个或多个上下文连接到它，系统就会保持 Service 运行。一旦这两种情况下 BIND_AUTO_CREATE 都不存在，就调用 Service 的 onDestroy()方法并终止 Service。所有的清理（停止线程，注销接收器）应该在从 onDestroy()方法返回时完成。

1．设置界面控件

在工程 ServiceApp 中，想启动 Service，需要设置界面控件，如【文件 9-3】所示。

【文件 9-3】 activity_main.xml

```
<?xml version = "1.0" encoding = "utf - 8"?>
< RelativeLayout xmlns:android = "http://schemas.android.com/apk/res/android"
    xmlns:app = "http://schemas.android.com/apk/res - auto"
    xmlns:tools = "http://schemas.android.com/tools"
    android:layout_width = "match_parent"
    android:layout_height = "match_parent"
    tools:context = ".MainActivity">
    < Button
        android:id = "@ + id/btn_startService"
        android:layout_width = "match_parent"
        android:layout_height = "wrap_content"
```

```xml
            android:layout_alignParentTop = "false"
            android:layout_centerHorizontal = "true"
            android:text = "启动服务(startService)" />
    < Button
            android:id = "@ + id/btn_stopService"
            android:layout_width = "match_parent"
            android:layout_height = "wrap_content"
            android:layout_below = "@id/btn_startService"
            android:text = "停止服务(stopService)" />
</RelativeLayout >
```

在【文件 9-3】中定义了 activity_main 的布局方式为相对布局,其中有两个按钮,id 分别为 btn_startService 和 btn_stopService,分别用来启动 Service 和停止 Service。

2. 编写界面交互程序

在 MainActivity 中实现对两个按钮的点击事件的监听,具体代码如【文件 9-4】所示。

【文件 9-4】 MainActivity.java

```java
package com.example.serviceapp;
import androidx.appcompat.app.AppCompatActivity;
import android.content.Intent;
import android.os.Bundle;
import android.view.View;
public class MainActivity extends AppCompatActivity implements View.OnClickListener{
    @Override
    protected void onCreate(Bundle savedInstanceState) {
        super.onCreate(savedInstanceState);
        setContentView(R.layout.activity_main);
        findViewById(R.id.btn_startService).setOnClickListener(this);
        findViewById(R.id.btn_stopService).setOnClickListener(this);
    }
    @Override
    public void onClick(View v) {
        Intent intent = new Intent(this, MyService.class);
        if (v.getId() == R.id.btn_startService) {
            //开启 Service
            startService(intent);
        }else if(v.getId() == R.id.btn_stopService){
            //停止 Service
            stopService(intent);
        }
    }
}
```

上述代码中,MainActivity 实现了 View.OnClickListener 接口。其中在 Override 的 onClick()方法中定义了 intent,指定 Context 为 this(MainActivity)以及要启动的 Service 为 MyService。之后通过一个 if 语句进行判定,如果单击的按钮为 btn_startService 则调用 startService()方法启动 Service,如果为 btn_stopService 则调用 stopService()方法停止 Service。

3. 重写 MyService 服务

对 MyService 类的实现进行重写,具体代码如【文件 9-5】所示。

【文件 9-5】 MyService.java

```java
package com.example.serviceapp;
……//省略 import 包(类)
public class MyService extends Service {
    public MyService() {
        Log.i("MyService","调用 MyService()构造方法");
    }
    @Override
    public void onCreate() {
        super.onCreate();
        Log.i("MyService","调用 onCreate()方法");
    }
    @Override
    public int onStartCommand(Intent intent, int flags, int startId) {
        Log.i("MyService","调用 onStartCommand()方法,intent = "
                + intent + "\tflags = " + flags + "\tstartId = " + startId);
        return super.onStartCommand(intent, flags, startId);
    }
    @Override
    public void onDestroy() {
        super.onDestroy();
        Log.i("MyService","调用 onDestroy()方法");
    }
    @Override
    public IBinder onBind(Intent intent) {
        Log.i("MyService","onBind()方法");
        return null;
    }
}
```

上述代码中对 Service 的构造方法、onCreate()方法、onStartCommand()方法、onDestroy()、onBind()方法进行了重写。当 Android 程序调用到 MyService 的不同生命周期时,会调用不同的方法,输出对应的 Log 信息。

在文件 9-4 的 MainActivity.java 中可以看到示例程序中使用 startService()方法启动 Service,根据 Service 的生命周期可以知道 onCreate()方法会被调用,然后会调用 onStartCommand()方法执行 Service 的逻辑操作。

在这里 onStartCommand()方法允许传递参数给 Service,以便 Service 在启动时能够获取这些参数并执行相应的操作。onStartCommand()方法中包含 3 个参数,其语法格式如下:

```
int onStartCommand(Intent intent, int flags, int startId)
```

其参数如下:

- intent:这是启动 Service 时传递的 Intent 对象。Intent 可以携带数据,例如操作标志、额外参数等。可以从这个 Intent 中提取数据,以便在 Service 中使用。
- flags:这是一个整数标志,表示启动 Service 的方式。它可以是以下值之一: START_FLAG_REDELIVERY、START_FLAG_RETRY、其他标志。
- startId:这是每次启动 Service 时唯一的标识符。每次启动 Service,startId 都会递增,使得可以在多次启动中区分不同的请求。

onStartCommand()方法的返回值表示系统在停止 Service 后的行为。它可以是以下值之一:START_NOT_STICKY、START_STICKY、START_REDELIVER_INTENT 等。在实际编程中,需要根据 Service 的性质和需求选择适当的返回值,不同的返回值会影响系统在 Service 异常终止后的行为。

如果调用 bindService()方法启动 Service,则 onCreate()方法被调用后,onBind()方法会被调用执行之后的逻辑操作。这里要注意,使用 bindService()方法启动 Service 多用于 AIDL 方式编程。

AIDL(Android Interface Definition Language)是 Android 接口定义语言,它允许不同的 Android 应用程序在不同的进程之间进行通信。AIDL 允许定义一个接口,使得一个进程中的组件可以调用另一个进程中的方法,就像调用本地方法一样。

4. 运行程序

运行上述程序,程序运行界面如图 9.4 所示,点击界面上的"启动服务(startService)"按钮,在 LogCat 窗口中会打印出开启 Service 过程中调用 Service 的生命周期方法,其中涉及 MyService 的构造方法、onCreate()方法、onStartCommand()方法,LogCat 窗口中输出的信息如图 9.5 所示。

图 9.4　ServiceApp 运行界面

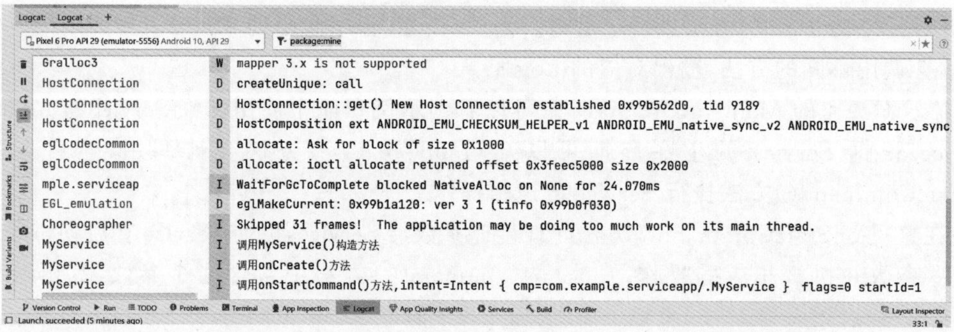

图 9.5　第一次点击"启动服务(startService)"按钮 LogCat 窗口中输出的信息

当再次点击"启动服务(startService)"按钮后,此时在 LogCat 窗口中会再次打印输出调用了 onStartCommand()方法,同时 startId 增加为 2,而构造方法和 onCreate()方法不再被调用,LogCat 窗口中输出的信息如图 9.6 所示。

当点击界面上的"停止服务(stopService)"按钮后,在 LogCat 窗口中打印出关闭 Service 过程中程序调用 MyService 的 onDestroy()生命周期方法,停止了 Service,LogCat 中输出的信息如图 9.7 所示。

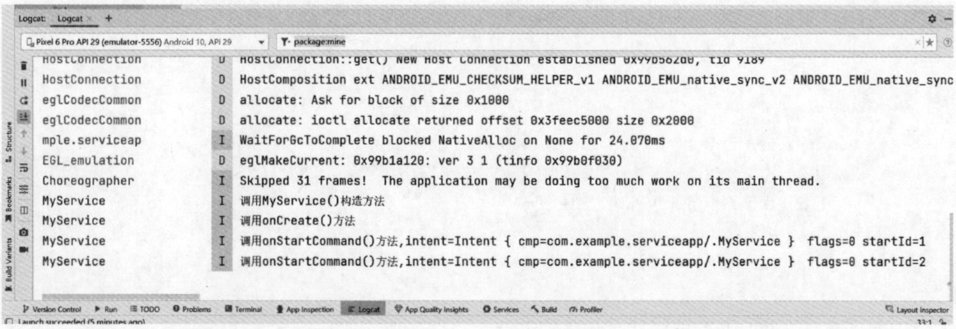

图9.6 第二次点击"启动服务(startService)"按钮 LogCat 窗口中输出的信息

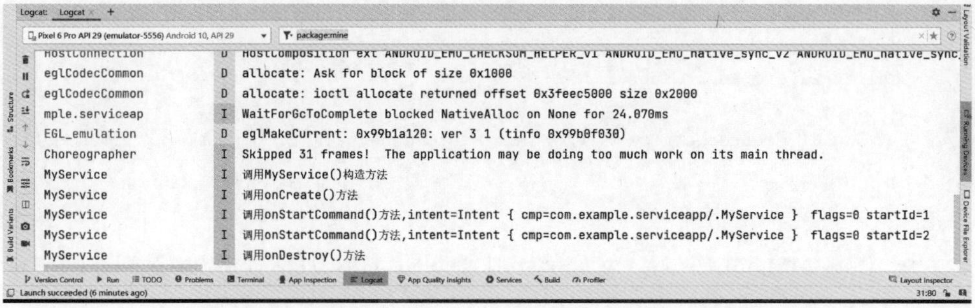

图9.7 点击"停止服务(stopService)"按钮 LogCat 窗口中输出的信息

9.3.3 绑定 Service

观看视频

启动 Service 有两种方式：一种是如 9.3.2 节所示调用 Context.startService()方法；另一种是调用 Context.bindService()方法绑定 Service,使用这种方式,Service 会绑定到调用者,并且只有调用者与 Service 处于绑定状态时,服务才会一直运行。当所有绑定的客户端(通常是 Activity)断开连接时,Service 会被销毁。

以下是一个简单的 Android 绑定 Service 的示例程序。在这个示例中,将创建一个计数器 Service,允许 Activity 绑定并获取计数器的当前值。

(1) 创建一个名为 CounterService 的服务类,具体代码如【文件 9-6】所示。

【文件 9-6】 CounterService.java

```
public class CounterService extends Service {

    private final IBinder binder = new LocalBinder();
    private int counter = 0;

    public class LocalBinder extends Binder {
        CounterService getService() {
            return CounterService.this;
        }
    }
    @Override
    public IBinder onBind(Intent intent) {
        Log.i("CounterService","调用 onBind()方法");
```

```java
        return binder;
    }

    @Override
    public boolean onUnbind(Intent intent) {
        Log.i("CounterService","调用 onUnbind()方法");
        return super.onUnbind(intent);
    }

    public int getCounter() {
        Log.i("CounterService","调用 getCounter()方法,获得 counter:" + counter);
        return counter;
    }

    public void incrementCounter() {
        counter++;
        Log.i("CounterService","调用 incrementCounter()方法,counter++:" + counter);
    }
}
```

(2) 在 AndroidManifest.xml 文件中注册这个 Service,具体代码如【文件 9-7】所示。

【文件 9-7】 AndroidManifest.xml

```xml
<manifest xmlns:android="http://schemas.android.com/apk/res/android"
    package="com.example.bindserviceexample">
    <application
        ……>
        <service android:name=".CounterService" />
        ……
    </application>
</manifest>
```

(3) 创建布局文件 activity_main.xml,具体代码如【文件 9-8】所示。

【文件 9-8】 activity_main.xml

```xml
<?xml version="1.0" encoding="utf-8"?>
<LinearLayout xmlns:android="http://schemas.android.com/apk/res/android"
    xmlns:app="http://schemas.android.com/apk/res-auto"
    xmlns:tools="http://schemas.android.com/tools"
    android:layout_width="match_parent"
    android:layout_height="match_parent"
    tools:context=".MainActivity"
    android:orientation="vertical">
    <Button
        android:id="@+id/btn_count"
        android:layout_width="match_parent"
        android:layout_height="wrap_content"
        android:text="调用绑定服务方法 incrementCounter()" />
</LinearLayout>
```

(4) 创建一个用于与服务交互的 Activity,具体代码如【文件 9-9】所示。

【文件 9-9】 MainActivity.java

```java
public class MainActivity extends AppCompatActivity {
    private boolean isBound = false;
    private CounterService counterService;

    private ServiceConnection connection = new ServiceConnection() {
        @Override
        public void onServiceConnected(ComponentName componentName, IBinder iBinder) {
            CounterService.LocalBinder binder = (CounterService.LocalBinder) iBinder;
            counterService = binder.getService();
            isBound = true;
        }

        @Override
        public void onServiceDisconnected(ComponentName componentName) {
            isBound = false;
        }
    };

    @Override
    protected void onCreate(Bundle savedInstanceState) {
        super.onCreate(savedInstanceState);
        setContentView(R.layout.activity_main);

        Intent intent = new Intent(this, CounterService.class);
        bindService(intent, connection, Context.BIND_AUTO_CREATE);
    }

    @Override
    protected void onDestroy() {
        super.onDestroy();
        if (isBound) {
            unbindService(connection);
        }
    }
}
```

在这个示例中，创建了 CounterService，这个 Service 有一个计数器。在 MainActivity 中 onCreate()方法中调用 bindService()方法绑定了这个 Service，并在 onServiceConnected()回调中获取了 Service 实例，当程序运行后点击 btn_count 按钮后，可以调用 Service 中的方法来获取计数器的值和递增计数器，若退出当前 Activity 则调用 onUnbind()方法。完整项目在 CountServiceApp 中，程序运行过程中，LogCat 窗口中的输出信息如图 9.8 所示。

图 9.8 CountServiceApp 运行过程中 LogCat 窗口中的输出信息

在9.3节中展示调用startService()和bindService()方法均为简单示例,在实际开发中可以根据实际需求进行扩展和修改。在实际应用中,可能需要更复杂的逻辑来处理绑定和解绑Service时的状态,图9.9所示为使用Service进行相关的异步操作通用时序图,仅供读者参考。

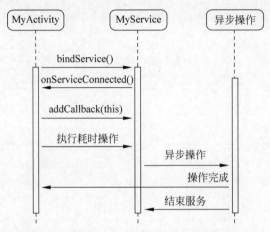

图9.9 使用Service进行异步操作通用时序图

观看视频

9.4 应用实例:音乐播放器

本章最后以一个简单的音乐播放器演示Service的使用方式,最终效果如图9.10所示。

(1)创建一个名为MusicServiceApp的应用程序,指定包名为cn.example.musicserviceapp,将本地音乐文件放入raw文件夹下(需要手动右击创建res下的raw目录,将相关音乐文件复制进去,这里复制进去的为canon.mp3),之后设计用户交互界面。程序主界面对应的布局具体代码详见【文件9-10】。

【文件9-10】 activity_main.xml

```
<?xml version = "1.0" encoding = "utf-8"?>
<RelativeLayout xmlns:android = "http://schemas.android.com/apk/res/android"
    xmlns:app = "http://schemas.android.com/apk/res-auto"
    xmlns:tools = "http://schemas.android.com/tools"
    android:layout_width = "match_parent"
    android:layout_height = "match_parent"
    android:background = "@drawable/bg"
    tools:context = ".MainActivity" >

    <Button
        android:id = "@+id/button"
        android:layout_width = "94dp"
        android:layout_height = "48dp"
        android:layout_alignParentStart = "true"
        android:layout_alignParentBottom = "true"
        android:background = "@drawable/play" />
```

图9.10 音乐播放器最终效果

```xml
<ProgressBar
    android:id = "@+id/progressBar"
    style = "?android:attr/progressBarStyleHorizontal"
    android:layout_width = "match_parent"
    android:layout_height = "wrap_content"
    android:layout_alignParentStart = "true"
    android:layout_alignParentTop = "true"
    android:progressTint = "@color/blue"
    android:layout_marginTop = "70dp" />
</RelativeLayout>
```

在 activity_main.xml 文件中，设置布局为 RelativeLayout，背景为 bg.png，放置了两个控件，一个是按钮（Button），另一个是进度条（ProgressBar）。这里要注意两个控件的相关设置，按钮使用 android:background 定义的按钮背景为 play.png，进度条设置 style 属性为水平进度条（?android:attr/progressBarStyleHorizontal）以及进度条颜色为蓝色（android:progressTint="@color/blue"）。

在程序运行过程中，由于 Android Studio 版本的不同，可能会出现按钮的背景图片无法显示的问题，这是要修改 themes.xml 文件中的 style 标签的 parent 属性设置，这里要修改为 parent="Theme.MaterialComponents.DayNight.DarkActionBar.Bridge"。设置如下：

```xml
<!-- style name = "Base.Theme.MusicServiceApp"
        parent = "Theme.Material3.DayNight.NoActionBar" -->
<style name = "Base.Theme.MusicServiceApp"
        parent = "Theme.MaterialComponents.DayNight.DarkActionBar.Bridge">
```

（2）创建 MusicService。

参考图 9.9，通过 Service 来做耗时的异步操作，音乐的加载、播放、网络文件的下载等都推荐使用 Service 操作。在这里，选中 com.example.musicserviceapp 的源代码包，右击后选择 New → Service → Service 选项，创建 MusicService。在该 Service 中，定义了 MusicBinder 内部类来和 MainActivity 进行本地通信，具体代码详见【文件 9-11】。

【文件 9-11】 MusicService.java

```java
package com.example.musicserviceapp;
import android.annotation.SuppressLint;
import android.app.Service;
import android.content.Intent;
import android.media.AudioManager;
import android.media.MediaPlayer;
import android.os.Binder;
import android.os.IBinder;

public class MusicService extends Service {
    private MediaPlayer mediaPlayer;
    public MusicService() {
    }

    @Override
```

```java
            public IBinder onBind(Intent intent) {
                //返回 MusicBinder 实例对象,与 MainActivity 进行本地通信
                return new MusicBinder();
            }
            class MusicBinder extends Binder
            {
                public void musicPlay()
                {
                    if(mediaPlayer == null)
                    {
                        mediaPlayer = MediaPlayer.create(MusicService.this,R.raw.canon);
                        mediaPlayer.setAudioStreamType(AudioManager.STREAM_MUSIC);
                        mediaPlayer.start();
                    }else
                    {
                        mediaPlayer.start();
                    }
                }

                public void musicPause()
                {
                    if(mediaPlayer!= null)
                    {
                        mediaPlayer.pause();
                    }
                }

                public void musicStop()
                {
                    if(mediaPlayer!= null)
                    {
                        mediaPlayer.stop();
                    }
                }

                public int getProgress()
                {
                    if(mediaPlayer!= null) {
                        return mediaPlayer.getCurrentPosition() * 100 / mediaPlayer.getDuration();
                    }else return 0;
                }

                @SuppressLint("SuspiciousIndentation")
                public void seekProgress(int progress)
                {
                    if(mediaPlayer!= null)
                        mediaPlayer.seekTo(progress * mediaPlayer.getDuration()/100);
                }
            }
        }
```

结合 Service 的生命周期以及 MusicService 类 onBind()方法中的代码可以看出,当 MainActivity 调用 bindService()方法启动 Service 后,onBind()方法会返回 MusicService 中 MusicBinder 内部类的实例给 MainActivity,通过这种方式,实现了 Activity 和 Service 的通信。

同时根据代码可以看出 MusicBinder 类为 Binder 类的子类，定义了音乐播放 musicPlay()、音乐暂停 musicPause()、音乐停止 musicStop()、得到进度 getProgress()、进度条跳转 seekProgress(int progress)这 5 个方法，通过这些方法来控制对 canon.mp3 音乐文件的播放控制。

（3）创建编写界面交互程序。

在 MainActivity.java 中，通过 bindService()方法启动 Service，具体代码详见【文件 9-12】。

【文件 9-12】 MainActivity.java

```
package com.example.musicserviceapp;
……//省略导入的相关包、类
public class MainActivity extends AppCompatActivity {
    private Button button;
    private ProgressBar progressBar;
    private MusicService.MusicBinder musicBinder;
    private boolean isPlay = false;
    private Timer timer;
    private TimerTask timerTask;
    private boolean isTimer = false;
    private TextView textView;

    @Override
    protected void onCreate(Bundle savedInstanceState) {
        super.onCreate(savedInstanceState);
        setContentView(R.layout.activity_main);
        button = findViewById(R.id.button);
        progressBar = findViewById(R.id.progressBar);
        Intent intent = new Intent(MainActivity.this,MusicService.class);
        bindService(intent, new ServiceConnection() {
            @Override
            public void onServiceConnected(ComponentName componentName, IBinder iBinder) {
                musicBinder = (MusicService.MusicBinder) iBinder;
            }

            @Override
            public void onServiceDisconnected(ComponentName componentName) {
            }
        }, BIND_AUTO_CREATE);
        timer = new Timer();
        timerTask = new TimerTask() {
            @Override
            public void run() {
                progressBar.setProgress(musicBinder.getProgress());
            }
        };

        button.setOnClickListener(new View.OnClickListener() {
            @Override
            public void onClick(View view) {
                if(isPlay == false)
                {
                    musicBinder.musicPlay();
                    Log.i("MainActivity","播放");
                    isPlay = true;
```

```
                    button.setBackgroundResource(R.drawable.pause);
                    if(isTimer == false)
                    {
                        timer.schedule(timerTask,100,20);
                        isTimer = true;
                    }
                    return;
                }else
                {
                    musicBinder.musicPause();
                    Log.i("MainActivity","暂停");

                    isPlay = false;

                    button.setBackgroundResource(R.drawable.play);
                    return;
                }
            }
        });
    }
}
```

MainActivity.java 中声明了 7 个属性，分别为按钮 button、进度条 progressBar、MusicService 内部类 MusicBinder 的引用 musicBinder、计时器 timer、时间任务 timerTask、判断音乐是否播放的布尔值 isPlay 和判断时间任务是否开启的布尔值 isTimer。

在 MainActivity 的 onCreate()方法中，调用 bindService()方法启动了 MusicService，bindService()方法的第二个参数定义了一个匿名内部类 ServiceConnection 对象，通过回调方法 onServiceConnected(ComponentName componentName, IBinder iBinder)返回 MusicBinder 的实例来进行 Activity 与 Service 的通信。之后通过 button 的 setOnClickListener()方法设置音乐的播放与暂停。程序运行结果及 LogCat 窗口中的输出信息如图 9.11 和图 9.12 所示。

图 9.11　音乐播放与暂停效果

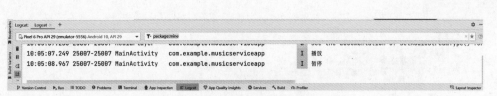

图 9.12　LogCat 窗口中的输出信息

9.5　小结

Service（服务）是一个功能强大的组件，使用时很容易变得非常复杂。在本章介绍了 Service 的基本概念、启动 Service 的两种方法，以及对应生命周期中方法的回调。最后，介绍了一个音乐播放器的实例。

习题 9

1. 简述不同启动方式下，Service 生命周期过程中所调用的回调方法有哪些。
2. 编写一个通过 Service，后台通过网络下载文件的程序。
3. 完善 9.4 节中音乐播放器的功能。

第10章

网络编程

 本章导图

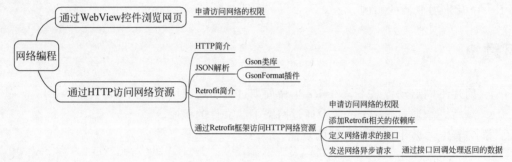

 主要内容

- 掌握 WebView 控件的使用，能够使用 WebView 浏览网页。
- 了解 HTTP，能够通过 HTTP 访问网络。
- 掌握 JSON 数据的解析，可以解析不同结构的 JSON 数据。
- 掌握 Retrofit 框架的应用，能够通过 Retrofit 框架完成 HTTP 请求。

 难点

- JSON 解析。
- Retrofit 框架的 HTTP 异步请求。

在移动互联网时代，手机联网实现信息互通是最基本的功能体验。上下班途中、休息旅行时，只要有空，人们就会拿出手机上网，通过手机接收新资讯、搜索网络资源。Android 作为智能手机市场主流的操作系统，它的强大离不开其对网络功能的支持。Android 系统提供了多种实现网络通信的方式，接下来，我们从基本的 WebView 控件开始，到 Android 通过 HTTP 进行网络通信以及网络数据的解析，详细讲解 Android 网络编程的相关知识。

10.1 通过 WebView 控件浏览网页

WebView 是一个能够显示网页的控件。但是它不是一个普通的 View 控件，它比一般的 View 控件要庞大、复杂，其实它就是一个浏览器，相当于一个 Google Chrome，当然它没有 Google Chrome 本身那么强大。它允许使用浏览器加载网页或者就在 WebView 内加载网页。早期的 WebView 控件使用 WebKit 渲染引擎来显示网页，在 Android 4.4 之后直接是基于 Chrome，然后 Chrome 又是基于 Chromium 内核。

如果要将 Web 应用程序（或只是网页）作为客户端应用程序的一部分提供，可以使用 WebView 进行操作。WebView 类是 Android 的 View 类的子类，它允许将网页显示为 Activity 布局的一部分。它不包括完全开发的 Web 浏览器的任何功能，例如导航控件或地址栏。默认情况下，所有 WebView 都会显示一个网页。

在开发原生 Android 应用过程中，为了赋予应用适当的动态性经常需要使用 WebView 控件在应用中内置 Web 网页，对于混合 App（Hybrid App）来说 WebView 控件更是必不可少，所以学习 WebView 控件的相关知识，对 WebView 控件有一个充分全面的了解对 Android 开发者来说非常重要。

WebView 控件是 Android 系统提供能显示网页的系统控件，它是一个特殊的 View，同时它也是一个 ViewGroup，可以有很多其他子 View。WebView 控件用来在应用中显示网页，就好比在应用中嵌入了一个浏览器窗口。有了这个浏览器窗口，还可以实现与 HTML 5 混合式开发，从而使应用具有跨平台、便于更新等优点。WebView 控件的用法和 ImageView 等控件的用法基本一样，使用时可以分两个步骤：第一步先把 WebView 控件添加到界面中；第二步通过调用某个方法载入网页。下面是一个简单的例子。

1. 创建应用程序

使用 Android Studio 新建 WebViewApp 应用程序，指定包名为 com.example.webviewapp。

2. 放置界面控件

在布局文件 activity_main.xml 中将页面布局改为 LinearLayout，方向设置为垂直，依次添加文本控件 EditText、按钮控件 Button 和 WebView 控件。其布局文件代码如下：

```
<?xml version = "1.0" encoding = "utf-8"?>
<LinearLayout xmlns:android = "http://schemas.android.com/apk/res/android"
    xmlns:app = "http://schemas.android.com/apk/res-auto"
    xmlns:tools = "http://schemas.android.com/tools"
    android:layout_width = "match_parent"
    android:layout_height = "match_parent"
    android:orientation = "vertical"
    tools:context = ".MainActivity">

    <EditText
        android:id = "@+id/editText"
        android:layout_width = "match_parent"
        android:layout_height = "wrap_content"
        android:ems = "10"
        android:hint = "网址"
        android:inputType = "text" />

    <Button
        android:id = "@+id/button"
        android:layout_width = "match_parent"
        android:layout_height = "wrap_content"
```

```
        android:text = "Button" />

    <WebView
        android:id = "@+id/wevView"
        android:layout_width = "match_parent"
        android:layout_height = "match_parent" />
</LinearLayout>
```

3. 在清单文件中添加访问网络的权限

打开 AndroidManifest.xml 文件,在 manifest 下添加申请访问网络的如下 3 个标记。

```
<uses-permission android:name = "android.permission.ACCESS_WIFI_STATE" />
<uses-permission android:name = "android.permission.INTERNET" />
<uses-permission android:name = "android.permission.ACCESS_NETWORK_STATE" />
```

4. 编写界面交互代码

在 MainActivity 中首先定义布局界面中的 3 个控件对象,具体代码如下:

```
private EditText editText;
private Button button;
private WebView webView;
```

在 onCreate()方法中依次进行初始化,初始化时设置允许 JavaScript 语言,因为现在大部分网页都使用了 JavaScript 语言,不允许 JavaScript 语言的话显示不完全。只有设置 WebView 控件的 WebViewClient 属性,才能将网页载入锁定在此控件中。不设置的话载入网页时会另外打开浏览器应用,而不会将网页载入 WebView 控件中。

```
editText = findViewById(R.id.editText);
button = findViewById(R.id.button);
webView = findViewById(R.id.wevView);
webView.getSettings().setJavaScriptEnabled(true);
webView.getSettings().setDomStorageEnabled(true);
webView.getSettings().setLoadWithOverviewMode(true);
webView.getSettings().setUseWideViewPort(true);
webView.getSettings().setBuiltInZoomControls(true);
webView.setWebViewClient(new WebViewClient());
```

为 Button 添加点击事件的监听器并实现,通过调用 WebView 的 loadUrl()方法,会加载参数对应网址的网页内容。

```
button.setOnClickListener(new View.OnClickListener() {
    @Override
    public void onClick(View v) {
        webView.loadUrl(editText.getText().toString());
    }
});
```

运行 WebViewApp 应用程序,在文本框中输入完整的 URL 网址 http://www.baidu.com,点击按钮就会在 WebView 控件中显示百度的首页内容,如图 10.1 所示。

在开发过程中需要注意,从 Android 9.0(API 级别为 28)开始,Android 默认将禁止明

图 10.1　WebView 控件浏览网页

文访问网络,只允许使用 HTTPS(超文本传输安全协议)访问指定的 URL 网页信息。如果要以明文的方式通过 HTTP 进行访问就需要进行 Android 网络安全配置。在本例中,需要在应用程序的资源文件夹下的 xml 文件夹下建立一个 network_security_config.xml 文件,该文件内容如下:

```
<?xml version = "1.0" encoding = "utf-8"?>
<network-security-config>
    <base-config cleartextTrafficPermitted = "true" />
</network-security-config>
```

在 AndroidManifest.xml 文件中增加 application 标记的属性,android:networkSecurityConfig=
"@xml/network_security_config",将其指向刚刚创建的网络安全配置文件 network_security_config.xml,如此一来就能够使用 HTTP,以明文的方式访问指定的 URL 网页内容了。在日常开发过程中,一般并不会配置 HTTPS 的 Web 服务器,故在开发过程中一般需要设置网络安全配置文件通过 HTTP 进行明文访问,待开发测试好后真正部署到支持 HTTPS 的 Web 服务器上时,再将明文访问的网络安全配置文件去除。本书后续的例子均需设置明文访问的网络安全配置文件,后面不再赘述。

10.2 通过 HTTP 访问网络资源

10.2.1 HTTP 简介

HTTP(Hyper Text Transfer Protocol,超文本传输协议)是一个应用层的协议,使用相当广泛,例如我们常说在浏览器中输入网址打开网页,浏览器与后台服务器之间就用的是 HTTP,并且也常用于后端各个微服务之间的数据请求和通信。它是 Web 技术的基石,也是互联网的基础技术之一。HTTP 通过传输 HTML、CSS、JavaScript 等静态资源文件和 API 接口等动态资源文件,提供支撑服务器响应用户请求的基础。HTTP 在移动应用程序中也扮演着重要的角色。移动应用通常都需要与服务器进行数据交互,例如获取社交网络应用的最新动态,或者是获得在线商城的商品信息。HTTP 通过提供快速、安全、可靠的数据传输,保证了服务器和移动应用的高效交互。

HTTP 的优点如下:

(1) HTTP 有着简单的请求-响应数据模型,非常易于理解。

(2) 简单快速。客户端向服务器请求服务时,只需要传送请求方法和路径。具有无状态性。

(3) 基于 TCP,通信过程稳定可靠,并且对开发者透明,可以不关注数据是如何传输的。

(4) 应用层的协议,可以方便地实现跨网络传输,通过 nginx 等网关可以方便地实现跨网络转发。

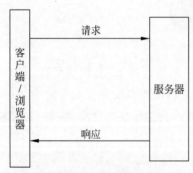

图 10.2　HTTP 请求响应模型

(5) 基于 HTTP,通常可以搭配 SSL 的数据加密技术,配合成为 HTTPS,保证数据通信过程的数据传输安全。

1. HTTP 请求响应模型

对于每一次的网络通信,一定会有发出的数据和返回的数据,在 HTTP 中就叫作请求(Request)和响应(Response)。例如通过浏览器访问网页,要查看的网页 URL、请求参数以及返回的网页内容等在 Response 对象中,如图 10.2 所示。

2. HTTP 请求方式

- OPTIONS:返回服务器针对特定资源所支持的 HTTP 请求方法。也可以利用向 Web 服务器发送'*'的请求来测试服务器的功能性。
- HEAD:向服务器索要与 GET 请求相一致的响应,只不过响应体将不会被返回。这一方法可以在不必传输整个响应内容的情况下,就可以获取包含在响应消息头中的元信息。
- GET:向特定的资源发出请求。
- POST:向指定资源提交数据进行处理请求(例如提交表单或者上传文件)。数据被包含在请求体中。POST 请求可能会导致新的资源的创建和/或已有资源的修改。
- PUT:向指定资源位置上传其最新内容。

- DELETE：请求服务器删除 Request-URI 所标识的资源。
- TRACE：回显服务器收到的请求，主要用于测试或诊断。
- CONNECT：HTTP 中预留给能够将连接改为管道方式的代理服务器。

3. 请求地址

HTTP 的请求地址就是通常说的网址，也即统一资源定位（Uniform Resource Locator，URL）。它一般包含以下几部分：

```
<请求协议>://<服务器名>:<端口>/<路径>
```

4. 请求数据

请求数据包括请求头部（Request Header）和请求正文（Request Body）。请求头部一般包含的是针对这次请求的附加信息，并不是请求的实际数据。这部分是针对数据的说明，有时也会携带 cookie 或者 token 等用于鉴权的信息。

请求正文是实际的数据部分。它的数据格式一般在请求头部中的 content-type 中进行说明，常见以下几种：

- application/json：JSON 数据格式。
- application/x-www-form-urlencoded：Ajax 方式发送默认的类型，form（表单）数据被编码为 key-value 格式发送到服务器（表单默认的提交数据的格式）。
- multipart/form-data：需要在表单中进行文件上传时，使用该格式。该格式下请求正文中会包含文件内容的起止位置说明信息。

10.2.2　JSON 解析

JSON（JavaScript Object Notation，JavaScript 对象简谱）是一种轻量级的数据交换格式。它基于 ECMAScript（欧洲计算机协会制定的 JavaScript 规范）的一个子集，采用完全独立于编程语言的文本格式来存储和表示数据。简洁和清晰的层次结构使得 JSON 成为理想的互联网数据交换格式。

JSON 是一个序列化的对象或数组。

对象由花括号括起来的以逗号分隔的成员构成，成员由字符串键和上文所述的值以逗号分隔的键值对组成，如：{"name":"John Doe","age":18,"address":{"country":"china","zip-code":"10000"}}。

Gson 是 Google 解析 JSON 的一个开源框架，它可以用来把 Java 对象转换为 JSON 表达式，也可以反过来把 JSON 字符串转换为与之相同的 Java 对象。

Gson 的目标：

- 提供简单易用的机制，类似于 toString()方法和构造器（工厂模式）用来进行 Java 和 JSON 互相转换。
- 允许把预先存在但无法修改的对象转换为 JSON 或从 JSON 中转换。
- 允许对象的自定义表示。
- 支持任何复杂的对象。
- 生成紧凑易读的 JSON 输出。

Gson 主要用到的类是 Gson，可以直接通过调用 new Gson()方法来生成，也可以用类

GsonBuilder 来创建 Gson 实例,这样创建可以自主进行参数设置,类似于版本控制。Gson 对象通过调用 toJson(obj)方法进行序列化,调用 fromJson(json,class)方法反序列化。反序列化时往往调用 obj.getClass()方法来获取类的信息。

下面的例子对上面的 JSON 字符串产生相应实体类的对象,并进行 JSON 数据格式的序列化与反序列化,再通过调用某个方法载入网页。

1. 创建应用程序

使用 Android Studio 新建 JSONApp 应用程序,指定包名为 com.example.jsonapp。

2. 放置界面控件

在布局文件 activity_main.xml 中将页面布局改为 LinearLayout,方向设置为垂直,依次添加 Button 控件和两个 TextView 控件。

3. 为项目添加 Gson 类库的依赖

选择 File→Project Structure 选项,如图 10.3 所示。

在 Project Structure 对话框中选择 Dependencies 选项,为应用 app 添加依赖,点击 Add Dependency,

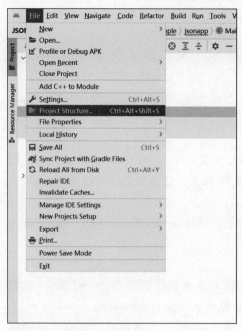

图 10.3 选择 File→Project Structure 选项

如图 10.4 所示。

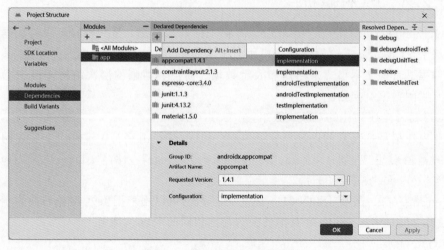

图 10.4 为 App 添加类库依赖

在弹出的对话框中输入要添加类库的名称 com.google.code.gson,点击 Search 按钮,选择 gson 2.10.1 后,点击 OK 按钮进行添加,如图 10.5 所示。

4. 添加 GsonFormat 插件

添加 gson 类库后,为了方便通过 JSON 字符串直接生成对应的实体类,可添加 GsonFormat 插件。首先选择 File→Settings 选项,如图 10.6 所示。

在 Settings 对话框中选择 Plugins 选项,然后在右侧的搜索栏中输入要安装的插件名

第10章 网络编程

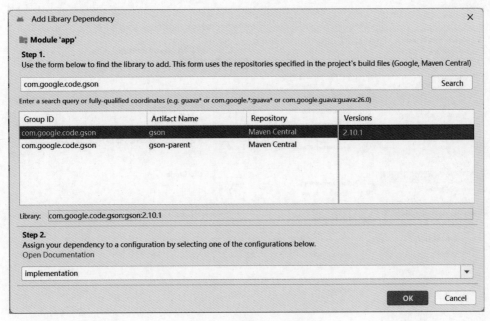

图 10.5　添加 gson 类库

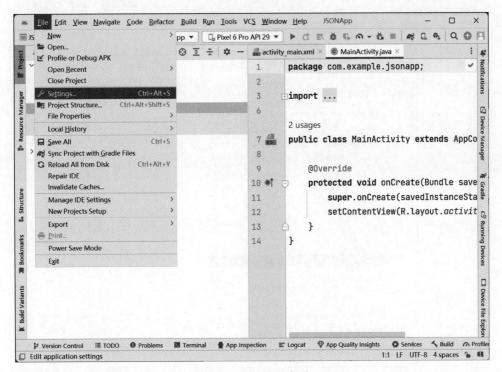

图 10.6　Settings 菜单

称 GsonFormat 进行搜索，搜索到后点击 install 按钮进行安装，如图 10.7 所示。

5．新建实体类 Person

在项目中新建 Java 类 Person，在新建的 Person 类中右击，选择 Generate 选项，如图 10.8 所示。

图 10.7　安装 GsonFormat 插件

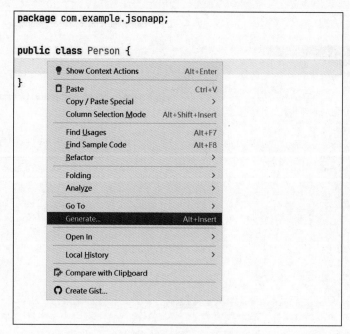

图 10.8　选择 Generate 选项

在弹出的菜单中选择 Gsonformat 选项,如图 10.9 所示。

在弹出的 GsonFormat 对话框中粘贴入完整的 JSON 字符串,本例输入 JSON 字符串:{"name":"John Doe","age":18,"address":{"country":"china","zip-code":"10000"}},如图 10.10 所示。

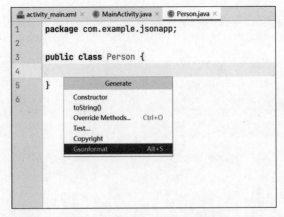

图 10.9　选择 Gsonformat 选项

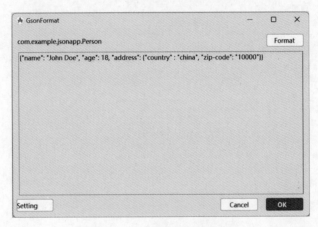

图 10.10　GsonFormat 对话框

点击 OK 按钮，选择所要生成的对应的实例类属性就可以了，如图 10.11 所示。

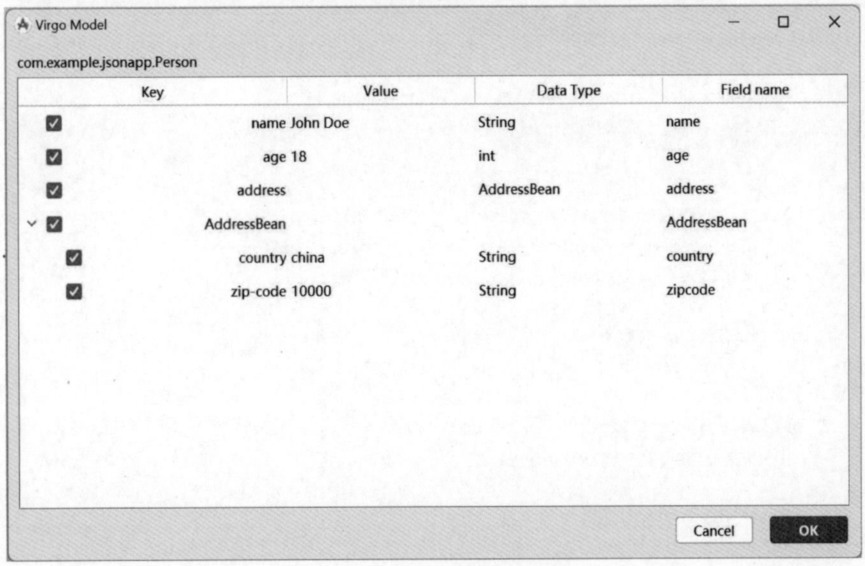

图 10.11　实体类模型

再次点击 OK 按钮即可生成 JSON 字符串所对应的实体类代码。所生成的 Person 类具体代码如下：

```java
public class Person {
    /**
     * name : John Doe
     * age : 18
     * address : {"country":"china","zip-code":"10000"}
     */

    private String name;
    private int age;
    private AddressBean address;

    public String getName() {
        return name;
    }
    public void setName(String name) {
        this.name = name;
    }
    public int getAge() {
        return age;
    }
    public void setAge(int age) {
        this.age = age;
    }
    public AddressBean getAddress() {
        return address;
    }
    public void setAddress(AddressBean address) {
        this.address = address;
    }

    public static class AddressBean {
        /**
         * country : china
         * zip-code : 10000
         */

        private String country;
        @com.google.gson.annotations.SerializedName("zip-code")
        private String zipcode;

        public String getCountry() {
            return country;
        }
        public void setCountry(String country) {
            this.country = country;
        }
        public String getZipcode() {
            return zipcode;
        }
        public void setZipcode(String zipcode) {
```

```
            this.zipcode = zipcode;
        }
    }
}
```

6. 实现 JSON 解析

在 MainActivity 中实现通过 JSON 字符串 jsonStr 生成对应的实体类对象 person，再由 person 对象转换为对应的 JSON 字符串，其具体的代码如下：

```java
public class MainActivity extends AppCompatActivity {
    private TextView textView1;
    private TextView textView2;
    private Button button;
    private Gson gson;
    private String jsonStr;
    private Person person;
    @Override
    protected void onCreate(Bundle savedInstanceState) {
        super.onCreate(savedInstanceState);
        setContentView(R.layout.activity_main);
        textView1 = findViewById(R.id.textView1);
        textView2 = findViewById(R.id.textView2);
        button = findViewById(R.id.button);
        gson = new Gson();
        jsonStr = "{\"name\": \"John Doe\", \"age\": 18, \"address\": {\"country\" : \"china\", \"zip-code\": \"10000\"}}";
        person = gson.fromJson(jsonStr,Person.class);
        button.setOnClickListener(new View.OnClickListener() {
            @Override
            public void onClick(View v) {
                textView1.setText(gson.toJson(person));
                textView2.setText(person.getName() + ":" + person.getAddress().getCountry());
            }
        });

    }
}
```

运行程序，点击 Button 按钮，在 TextView1 上可以看到 person 对象生成的对应的 JSON 字符串，在 TextView2 可以看到通过 JSON 字符串解析出的 person 对象的姓名和国别，如图 10.12 所示。

10.2.3 Retrofit 简介

Android 的应用程序如果想要通过 HTTP 访问互联网上的资源可以通过 URLConnection 类。URLConnection 类是 Java JDK 中自带的网络请求类，HttpURLConnection 继承自 URLConnection 类，其对象主要通过 URL 的 openConnection() 方法获得，能够采用 GET 或 POST 的方式请求 HTTP 的资源，通过网络流的方式来处理接收到的网络数据，在日常使用中并不是很方便。

OkHttp 是一个优秀的网络请求框架，由 Square 公司设计研发并开源，目前可以在

图 10.12　JSONApp 运行结果

Java 和 Kotlin 中使用；主要用于替代 HttpUrlConnection，提供了对 HTTP/2 和 SPDY 的支持，并提供了连接池、GZIP 压缩和 HTTP 响应缓存等功能。

　　Retrofit 也是由 Square 公司设计研发并开源的，是一个支持 RESTful 的 HTTP 网络请求框架的封装。Retrofit 2.0 开始内置了 OkHttp，Android 的应用程序通过 Retrofit 请求网络，实际上是使用 Retrofit 接口层封装请求参数、Header、URL 等信息，之后由 OkHttp 完成后续的请求操作，在服务端返回数据之后，OkHttp 将原始的结果交给 Retrofit，后者根据用户的需求对结果进行解析。

　　Retrofit 的优点：
- 超级解耦，接口定义、接口参数、接口回调不再耦合在一起。
- 可以配置不同的 httpClient 来实现网络请求，如 okHttp、httpClient。
- 支持同步、异步、Rxjava。
- 可以配置不同反序列化工具类来解析不同的数据，如 JSON、XML。
- 请求速度快，使用方便灵活简洁。

　　Retrofit 使用大量注解来简化请求，Retrofit 将 okHttp 请求抽象成 Java 接口，使用注解来配置和描述网络请求参数。

1. 请求方法注解

请求方法注解见表 10.1。

表 10.1　请求方法注解

请求方法注解	说　　明
@GET	GET 请求
@POST	POST 请求
@PUT	PUT 请求
@DELETE	DELETE 请求
@PATCH	PATCH 请求,该请求是对 PUT 请求的补充,用于更新局部资源
@HEAD	HEAD 请求
@OPTIONS	OPTIONS 请求
@HTTP	通过注解,可以替换以上所有的注解,它拥有三个属性:method、path、hasBody

2. 请求头注解

请求头注解见表 10.2。

表 10.2　请求头注解

请求头注解	说　　明
@Headers	用于添加固定请求头,可以同时添加多个,通过该注解的请求头不会相互覆盖,而是共同存在
@Header	作为方法的参数传入,用于添加不固定的 Header,它会更新已有请求头

3. 请求参数注解

请求参数注解见表 10.3。

表 10.3　请求参数注解

请求参数注解	说　　明
@Body	多用于 POST 请求发送非表达数据,根据转换方式将实例对象转换为对应字符串传递参数,如使用 POST 请求发送 JSON 数据,添加 GsonConverterFactory 则是将 body 转换为 JSON 字符串进行传递
@Filed	多用于 POST 方式传递参数,需要结合@FromUrlEncoded 使用,即以表单的形式传递参数
@FiledMap	多用于 POST 请求中的表单字段,需要结合@FromUrlEncoded 使用
@Part	用于表单字段,Part 和 PartMap 与@multipart 注解结合使用,适合文件上传的情况
@PartMap	用于表单字段,默认接受类型是 Map<String,RequestBody>,可用于实现多文件上传
@Path	用于 URL 中的占位符
@Query	用于 GET 请求中的参数
@QueryMap	与 aQuery 类似,用于不确定表单参数
@Url	指定请求路径

4. 请求和响应格式(标记)注解

请求和响应格式(标记)注解见表 10.4。

表 10.4 请求和响应格式(标记)注解

请求和响应格式(标记)注解	说　明
@FromUrlCoded	表示请求发送编码表单数据,每个键值对需要使用@Filed 注解
@Multipart	表示请求发送 form_encoded 数据(适用于有文件上传的场景),每个键值对需要用@Part 来注解键名,随后的对象需要提供值
@Streaming	表示响应用字节流的形式返回,如果没有使用注解,默认会把数据全部载入内存中,该注解在下载大文件时特别有用

10.2.4 通过 Retrofit 框架访问 HTTP 网络资源

下面通过一个例子来说明如何使用 Retrofit 框架访问 HTTP 网络资源。例如可以将一个 JSON 文件部署在一个 Web 服务器上,或者部署一个 Web 服务的应用程序能够返回 JSON 数据。在本例中将一个包含了天气预报中所使用的全国各地城市代码的 JSON 文件部署在 Web 服务器上,如访问地址为 https://rg.itsven.cn/static/json/City.json,该地址能够通过浏览器访问,其访问结果如图 10.13 所示。

图 10.13　全国各地城市代码的 JSON 数据

使用 Retrofit 框架可通过 HTTP 请求到该 JSON 数据,并查询到任一城市的天气预报城市代码。其具体实现步骤如下:

1. 创建应用程序

使用 Android Studio 新建 CidyCodeApp 应用程序,指定包名为 com.example.cidycodeapp。

2. 放置界面控件

在布局文件 activity_main.xml 中将页面布局改为 LinearLayout,方向设置为垂直,依次添加 EditText 控件、Button 控件和 ListView 控件。

3. 在清单文件中添加访问网络的权限

打开 AndroidManifest. xml 文件，在< manifest >标记下添加申请访问网络的如下三个标记。

```
< uses - permission android:name = "android.permission.ACCESS_WIFI_STATE" />
< uses - permission android:name = "android.permission.INTERNET" />
< uses - permission android:name = "android.permission.ACCESS_NETWORK_STATE" />
```

注意，如果 City. json 文件是部署在自己不支持 HTTPS 的 Web 服务器上，需仿照 10.1.1 节中的例子设置 network_security_config，以明文的方式访问指定的 URL 网络内容。

4. 添加 Retrofit 相关的依赖库

如果要使用 Retrofit 框架，必须添加相应的类库依赖。主要有表 10.5 中所示的依赖。

表 10.5 Retrofit 相关依赖

类库名称	Gradle 依赖
Gson	com. google. code. gson:gson:2. 10. 1
Retrofit2	com. squareup. retrofit2:retrofit:2. 9. 0
Retrofit2 Converter-gson	com. squareup. retrofit2:converter-gson:2. 0. 2

对于类库依赖，可仿照 10.2.2 节中的例子通过在 Project Structure 中一项一项添加，也可以直接在项目路径下 app 文件夹下的 build. gradle 文件中直接添加其依赖信息，如图 10.14 所示。

```
dependencies {

    implementation 'androidx.appcompat:appcompat:1.4.1'
    implementation 'com.google.android.material:material:1.5.0'
    implementation 'androidx.constraintlayout:constraintlayout:2.1.3'
    testImplementation 'junit:junit:4.13.2'
    androidTestImplementation 'androidx.test.ext:junit:1.1.3'
    androidTestImplementation 'androidx.test.espresso:espresso-core:3.4.0'
    implementation 'com.google.code.gson:gson:2.10.1'
    implementation 'com.squareup.retrofit2:retrofit:2.9.0'
    implementation 'com.squareup.retrofit2:converter-gson:2.0.2'

}
```

图 10.14 项目所需依赖

5. 创建服务器返回 JSON 数据对应的实体类

仿照 10.2.2 节中的例子生成 JSON 字符串对应的实体类。通过浏览器访问到全国各地城市代码的 JSON 数据，随便复制一个城市完整的 JSON 数据，使用 GsonFormat 插件生成城市代码 JSON 数据所对应的实体类 CityCode。其具体代码如下：

```
public class CityCode {
    /**
     * _id : 147
     * id : 148
```

```
             * pid : 10
             * city_code : 101180101
             * city_name : 郑州
             */
            private int _id;
            private int id;
            private int pid;
            private String city_code;
            private String city_name;
            public int get_id() {
                return _id;
            }
            public void set_id(int _id) {
                this._id = _id;
            }
            public int getId() {
                return id;
            }
            public void setId(int id) {
                this.id = id;
            }
            public int getPid() {
                return pid;
            }
            public void setPid(int pid) {
                this.pid = pid;
            }
            public String getCity_code() {
                return city_code;
            }
            public void setCity_code(String city_code) {
                this.city_code = city_code;
            }
            public String getCity_name() {
                return city_name;
            }
            public void setCity_name(String city_name) {
                this.city_name = city_name;
            }
        }
```

6. 定义用于描述网络请求的接口

由请求的 URL 地址可知,HTTP 请求的方式为 GET 方式,请求的具体文件为 City.json。可在项目中添加描述网络请求的接口 RequestCityCode。其具体代码如下:

```
public interface RequestCityCode {
    @GET("City.json")
    Call<List<CityCode>> getCityCodesAll();
},
```

这里@GET 注解表示要使用 GET 方式提交 HTTP 请求,请求的子路径为 City.json。请求时调用的方法 getCityCodesAll()将会返回包含一系列城市代码类对象的一个 List 集合。

7. 创建 Retrofit 实例

在 MainActivity 中要使用 Retrofit，首先要创建其对象，并设置网络请求的 URL 主路径以及使用 Gson 解析。其具体代码如下：

```
retrofit2 = new Retrofit.Builder()
            .baseUrl("https://rg.itsven.cn/static/json/")     //设置网络请求 URL
            .addConverterFactory(GsonConverterFactory.create())
                                        //设置使用 Gson 解析(记得加入依赖)
            .build();
```

8. 创建网络请求接口实例

Retrofit 框架会根据前面定义好的网络请求的接口自动生成所对应的类，这里需要创建出该类对象以便进行网络请求。其具体代码如下：

```
//创建网络请求接口的实例
RequestCityCode requestCityCode = retrofit2.create(RequestCityCode.class);
```

9. 发送网络请求（异步）

首先通过网络请求接口的实例调用其方法获得请求对象，然后通过请求对象发起异步请求。其具体代码如下：

```
Call<List<CityCode>> call = requestCityCode.getCityCodesAll();
        call.enqueue(new Callback<List<CityCode>>() {
            @Override
            public void onResponse(Call<List<CityCode>> call, Response<List<CityCode>> response) {

            }
            @Override
            public void onFailure(Call<List<CityCode>> call, Throwable t) {

            }
});
```

10. 处理返回的数据

异步请求中会生成两个接口回调的方法：一个是 onResponse()方法；另一个是 onFailure()方法。如果请求成功，则会自动调用 onResponse()方法，故需在该方法中进行获得到返回数据的处理；如果请求失败，则会自动调用 onFailure()方法，在该方法中可获取网络请求失败的具体信息。以下代码是网络请求成功 onResponse()方法中的代码：

```
@Override
            public void onResponse(Call<List<CityCode>> call, Response<List<CityCode>> response) {
                cityCodeList = response.body();
                for(CityCode cityCode:cityCodeList)
                {
cityList.add(cityCode.getCity_name() + ":" + cityCode.getCity_code());
                }
                arrayAdapter = new
ArrayAdapter<>(MainActivity.this,android.R.layout.simple_list_item_1,cityList);
                listView.setAdapter(arrayAdapter);
            }
```

将请求到的全国各地城市名称及代码显示在界面的 ListView 中。

若请求成功则获取了全国各地城市名称及代码,在文本框中输入某一城市的城市名称,点击按钮对请求到的数据集进行检索,可显示对应的城市代码。其最终程序运行结果如图 10.15 所示。

图 10.15　程序运行结果

附 MainActivity.java 文件的完整代码如下：

```
import androidx.appcompat.app.AlertDialog;
import androidx.appcompat.app.AppCompatActivity;

import android.os.Bundle;
import android.view.View;
import android.widget.ArrayAdapter;
import android.widget.Button;
import android.widget.EditText;
import android.widget.ListView;
```

```java
import java.util.ArrayList;
import java.util.List;

import retrofit2.Call;
import retrofit2.Callback;
import retrofit2.Response;
import retrofit2.Retrofit;
import retrofit2.converter.gson.GsonConverterFactory;

public class MainActivity extends AppCompatActivity {

    private EditText editText2;
    private Button button2;
    private RequestCityCode requestCityCode;

    private List<CityCode> cityCodeList;
    private String cityId;
    private Retrofit retrofit2;
    private List<String> cityList;
    private ArrayAdapter<String> arrayAdapter;
    private ListView listView;

    @Override
    protected void onCreate(Bundle savedInstanceState) {
        super.onCreate(savedInstanceState);
        setContentView(R.layout.activity_main);
        editText2 = findViewById(R.id.editText2);
        button2 = findViewById(R.id.button2);
        listView = findViewById(R.id.listView);
        //创建 Retrofit 对象
        retrofit2 = new Retrofit.Builder()
                .baseUrl("https://rg.itsven.cn/static/json/")    // 设置网络请求 URL
                .addConverterFactory(GsonConverterFactory.create())
                //设置使用 Gson 解析(记得加入依赖)
                .build();
        //网络请求接口的实例
        requestCityCode = retrofit2.create(RequestCityCode.class);
        getAllCityCodes();
        cityList = new ArrayList<>();
        button2.setOnClickListener(new View.OnClickListener() {
            @Override
            public void onClick(View view) {
                String cityID = "";
                for(CityCode cityCode:cityCodeList)
                {
if(cityCode.getCity_name().equals(editText2.getText().toString().trim()))
                    {
                        cityID = cityCode.getCity_code();
                            AlertDialog.Builder alertDialog = new AlertDialog.Builder(MainActivity.this);
                        alertDialog.setTitle(editText2.getText().toString().trim());
                        alertDialog.setMessage("Code:" + cityID);
                        alertDialog.setPositiveButton("确定",null);
```

```java
                    alertDialog.show();
                }
            }

        }
    });
}

private void getAllCityCodes()
{
    Call<List<CityCode>> call = requestCityCode.getCityCodesAll();
    call.enqueue(new Callback<List<CityCode>>() {
        @Override
        public void onResponse(Call<List<CityCode>> call, Response<List<CityCode>> response) {
            cityCodeList = response.body();
            for(CityCode cityCode:cityCodeList)
            {
             cityList.add(cityCode.getCity_name() + ":" + cityCode.getCity_code());
            }
             arrayAdapter = new ArrayAdapter<>(MainActivity.this,android.R.layout.simple_list_item_1,cityList);
            listView.setAdapter(arrayAdapter);
        }

        @Override
        public void onFailure(Call<List<CityCode>> call, Throwable t) {

        }
    });
}
}
```

10.3 应用实例：天气预报

通过本章的学习可以实现一个天气预报的 Android 应用程序。在 10.2.4 节中 CityCodeApp 例子的基础上，可查询出全国各个城市的代码，并通过城市代码查询出该城市的天气预报。这里就需要通过网络请求到某个城市的天气预报数据。目前在互联网中很多企业都会提供天气预报的数据，如果不是从事商业活动，仅仅用于开发测试，大多也都是免费的，如中国天气网、百度天气、和风天气、心知天气等。在网络请求过程中大体也基本相同，本实例中会使用和风天气网所提供的天气预报数据，需提前注册其账号，并通过订阅免费服务获取一个和风天气开发服务的密钥 KEY(或称为 Token，认证信息)。例如在浏览器中输入下列 API 地址(请将最后的**你的 KEY** 替换成你申请到的密钥 KEY)：

https://devapi.qweather.com/v7/weather/now?location=101010100&key=你的KEY

此时应该就可以获得北京市的实时天气数据了。这样就有了天气预报的数据。

利用本书第 6 章学习的内容，可以将经常访问的城市添加到手机自带的 SQLite 数据库中,便于访问常用城市的天气预报。

1. 创建应用程序

使用 Android Studio 新建 CidyCodeApp 应用程序,指定包名为 com.example.weatherapp。

2. 实现通过网络获取全国各地城市代码

首页 MainActivity 的实现与 10.2.4 节中的例子实现基本相同,在此不再赘述。界面上添加了一个"添加"按钮和一个"下一页"按钮。"添加"按钮可把查询到的城市添加到常用城市列表中,"下一页"按钮可跳转至常用城市页面,其显示效果如图 10.16 所示。

图 10.16　首页显示效果

3. 添加城市到常用城市列表

这部分实现过程与 6.4 节的实现过程基本相同,主要是将城市名称和查询到的城市代码保存到本地的 SQLite 数据库中。新建数据表的 SQL 语句如下:

```
CREATE TABLE information(id INTEGER PRIMARY KEY AUTOINCREMENT," +
            "name VARCHAR(20),cityCode VARCHAR(50) )
```

数据表名称为 information,该表有 3 个字段,其中 id 为主键,数据类型为整型,自动递增;name 字段为字符串类型,长度为 20,用于保存城市名称;cityCode 字段为字符串类型,长度为 50,用于保存城市代码。封装操作 information 表的数据访问类 CityDBHelper,其具体实现的代码如下:

```
import android.content.ContentValues;
import android.content.Context;
import android.database.Cursor;
import android.database.sqlite.SQLiteDatabase;
```

```java
import android.database.sqlite.SQLiteOpenHelper;
import android.util.Log;
import android.widget.Toast;
import androidx.annotation.Nullable;
import java.util.ArrayList;

public class CityDBHelper extends SQLiteOpenHelper {
    private Context context;

    public CityDBHelper(@Nullable Context context) {
        super(context, "rg1ContactDB", null, 1);
        this.context = context;
    }

    @Override
    public void onCreate(SQLiteDatabase sqLiteDatabase) {
        sqLiteDatabase.execSQL(" CREATE TABLE information (_id INTEGER PRIMARY KEY AUTOINCREMENT," + "name VARCHAR(20),cityCode VARCHAR(50) )");
    }

    @Override
    public void onUpgrade(SQLiteDatabase sqLiteDatabase, int i, int i1) {

    }

    public long insert(String name, String cityCode) {
        SQLiteDatabase db = this.getWritableDatabase();
        ContentValues values = new ContentValues();
        values.put("name", name);
        values.put("cityCode", cityCode);
        long id = db.insert("information", null, values);
        Toast.makeText(context, "插入成功!", Toast.LENGTH_SHORT).show();
        return id;
    }

    public String queryByName(String name) {
        SQLiteDatabase db = this.getReadableDatabase();
        Cursor cursor = db.query("information", null,
                "name = ?", new String[]{name}, null, null, "name");
        if (cursor.getCount() == 0) {
            cursor.close();
            db.close();
            return "NULL";
        } else {
            cursor.moveToFirst();
            String cityCode = cursor.getString(2);
            cursor.close();
            db.close();
            return cityCode;
        }
    }

    public String queryAll() {
```

```java
        SQLiteDatabase db = this.getReadableDatabase();
        Cursor cursor = db.query("information", null,
            null, null, null, null, "name");
        if (cursor.getCount() == 0) {
            cursor.close();
            db.close();
            return "NULL";
        } else {
            StringBuilder infos = new StringBuilder();
            while (cursor.moveToNext()) {
                infos.append("姓名:" + cursor.getString(1) + " , " + "手机:" + cursor.getString(2) + "\n");
            }
            cursor.close();
            db.close();
            return infos.toString();
        }
    }

    public ArrayList<City> getAllContacts() {
        ArrayList<City> cities = new ArrayList<City>();
        SQLiteDatabase db = this.getReadableDatabase();
        Cursor cursor = db.query("information", null, null, null, null,
            null, null);
        City city;
        while (cursor.moveToNext()) {
            city = new City(cursor.getInt(0), cursor.getString(1), cursor.getString(2));
            cities.add(city);
        }
        cursor.close();
        db.close();
        return cities;
    }
```

在文本框中输入城市名称(如：郑州)，点击"添加"按钮可将城市名称和查询到的城市代码保存到数据表中，在 MainActivity 中的具体实现如下：

```java
buttonAdd.setOnClickListener(new View.OnClickListener() {
    @Override
    public void onClick(View v) {
        String city = editText.getText().toString().trim();
        for(CityEntity cityEntity:cityList)
        {
            if(city.equals(cityEntity.getCity_name()))
            {
                cityDBHelper.insert(cityEntity.getCity_name(),cityEntity.getCity_code());
                Toast.makeText(MainActivity.this,city+ "添加成功！",Toast.LENGTH_SHORT).show();
            }
        }
    }
});
```

4. 实现常用城市列表页

点击首页上的"下一页"按钮,可跳转至常用城市列表页面。该页面主要由一个 ListView 控件构成,ListItem 列表项的布局主要由一个 ImageView 和两个 TextView 构成,ImageView 可以用来显示城市的图片,两个 TextView 分别显示程序名称和城市代码。常用城市列表页显示效果如图 10.17 所示。

图 10.17　常用城市列表页显示效果

常用城市列表页的实现首先需从数据表中读取保存的常用城市信息,通过 CityDBHelper 对象调用其 getAllContacts()方法,将保存的常用城市信息读取到 List < City >集合 cityList 中。CityArrayAdapter 继承 ArrayAdapter < CityItem >类来实现 ListView 的数据适配器类 CityArrayAdapter,ListView 控件 listViewCity 通过设置适配器对象 cityArrayAdapter 来完成对数据的显示。其适配器 CityArrayAdapter 类的具体代码如下:

```
import android.content.Context;
import android.view.LayoutInflater;
import android.view.View;
import android.view.ViewGroup;
import android.widget.ArrayAdapter;
import android.widget.ImageView;
import android.widget.TextView;
import androidx.annotation.NonNull;
import androidx.annotation.Nullable;
import com.bumptech.glide.Glide;
import java.util.List;

public class CityArrayAdapter extends ArrayAdapter< CityItem >
{
    /**
```

```java
 * Constructor
 *
 * @param context The current context.
 * @param resource The resource ID for a layout file containing a TextView to use when
 *                 instantiating views.
 * @param objects The objects to represent in the ListView.
 */
public CityArrayAdapter(@NonNull Context context, int resource, @NonNull List<CityItem> objects) {
    super(context, resource, objects);
}

@NonNull
@Override
public View getView(int position, @Nullable View convertView, @NonNull ViewGroup parent) {
    CityItem cityItem = getItem(position);
    View view = LayoutInflater.from(getContext()).inflate(R.layout.list_item,null);
    ImageView imageView = view.findViewById(R.id.imageView);
    TextView textViewCity = view.findViewById(R.id.textViewCity);
    TextView textViewCityID = view.findViewById(R.id.textViewCityID);
    textViewCity.setText(cityItem.getName());
    textViewCityID.setText(cityItem.getCityID());
    Glide.with(getContext()).load(cityItem.getImg()).into(imageView);
    return view;

}
}
```

常用城市列表页 CityActivity 类的具体实现代码如下：

```java
import androidx.appcompat.app.AppCompatActivity;
import android.content.Intent;
import android.os.Bundle;
import android.view.View;
import android.widget.AdapterView;
import android.widget.ArrayAdapter;
import android.widget.ListView;
import java.util.ArrayList;
import java.util.List;

public class CityActivity extends AppCompatActivity {
    private ListView listViewCity;
    private CityDBHelper cityDBHelper;
    private List<City> cityList;
    //private List<String> cityStrList;
    private ArrayAdapter<String> arrayAdapter;
    private List<CityItem> cityItemList;
    private int count = 0;
    private CityArrayAdapter cityArrayAdapter;

    @Override
    protected void onCreate(Bundle savedInstanceState) {
```

```java
        super.onCreate(savedInstanceState);
        setContentView(R.layout.activity_city);
        listViewCity = findViewById(R.id.listViewCity);
        cityDBHelper = new CityDBHelper(this);
        cityList = cityDBHelper.getAllContacts();
        //cityStrList = new ArrayList<>();
        cityItemList = new ArrayList<>();
        for(City city : cityList)
        {
//cityStrList.add(contact.getName() + ":\n" + contact.getcityCode());
            count = (count + 1) % 10;
            CityItem cityItem = new CityItem();
            cityItem.setName(city.getName());
            cityItem.setCityID(city.getcityCode());
            cityItem.setImg("https://rg.itsven.cn/static/img/picture" + count + ".jpg");
            cityItemList.add(cityItem);
        }
        //arrayAdapter = new ArrayAdapter<>(this,android.R.layout.simple_list_item_1,cityStrList);
        //listViewCity.setAdapter(arrayAdapter);
        cityArrayAdapter = new CityArrayAdapter(this,R.layout.list_item,cityItemList);
        listViewCity.setAdapter(cityArrayAdapter);
        listViewCity.setOnItemClickListener(new AdapterView.OnItemClickListener() {
            @Override
            public void onItemClick(AdapterView<?> parent, View view, int position, long id) {
                Intent intent = new Intent(CityActivity.this,WeatherActivity.class);
                intent.putExtra("cityName",cityItemList.get(position).getName());
                intent.putExtra("cityID",cityItemList.get(position).getCityID());
                startActivity(intent);
            }
        });
    }
}
```

点击某个城市的列表项可进入下一页显示该城市的天气预报信息。

5. 实现天气预报页面

首先在 AndroidManifest.xml 文件中申请访问网络的权限,在 manifest 下添加申请访问网络的如下三个标记。

```
< uses - permission android:name = "android.permission.ACCESS_WIFI_STATE" />
< uses - permission android:name = "android.permission.INTERNET" />
< uses - permission android:name = "android.permission.ACCESS_NETWORK_STATE" />
```

其次添加该应用程序所需的类库依赖,主要有表 10.6 所示的依赖。

表 10.6 程序所需的类库依赖

类库名称	Gradle 依赖
Gson	com.google.code.gson:gson:2.10.1
Retrofit2	com.squareup.retrofit2:retrofit:2.9.0

类 库 名 称	Gradle 依赖
Retrofit2 Converter-gson	com.squareup.retrofit2:converter-gson:2.0.2
Glide	com.github.bumptech.glide:glide:4.12.0

可以直接在项目路径下 app 文件夹下的 build.gradle 文件中直接添加其依赖信息,代码如下:

```
implementation 'com.google.code.gson:gson:2.10.1'
implementation 'com.squareup.retrofit2:retrofit:2.9.0'
implementation 'com.squareup.retrofit2:converter-gson:2.0.2'
implementation 'com.github.bumptech.glide:glide:4.12.0'
```

其中,glide:4.12.0 类库用于通过图片文件的 URL 为图像控件(如 ImageView 等)加载网络图片。

接下来通过浏览器请求某个城市的天气预报 JSON 数据,其请求地址如下:

```
https://devapi.qweather.com/v7/weather/now?location=101010100&key=你的KEY
```

这里需要注意请将最后的"你的 KEY"替换成你申请到的密钥 KEY,此时应该就可以获得北京市的实时天气 JSON 数据了,其 JSON 数据格式如下:

```
{
    "code": "200",
    "updateTime": "2023-08-17T18:17+08:00",
    "fxLink": "https://www.qweather.com/weather/beijing-101010100.html",
    "now": {
        "obsTime": "2023-08-17T18:13+08:00",
        "temp": "31",
        "feelsLike": "34",
        "icon": "104",
        "text": "阴",
        "wind360": "167",
        "windDir": "南风",
        "windScale": "1",
        "windSpeed": "2",
        "humidity": "64",
        "precip": "0.0",
        "pressure": "1000",
        "vis": "13",
        "cloud": "91",
        "dew": "26"
    },
    "refer": {
        "sources": ["QWeather", "NMC", "ECMWF"],
        "license": ["CC BY-SA 4.0"]
    }
}
```

根据该 JSON 数据格式通过 GsonFormat 插件得到对应的实体类 WeatherInfo,其具体代码如下:

```java
import java.util.List;

public class WeatherInfo {
    private String code;
    private String updateTime;
    private String fxLink;
    private NowBean now;
    private ReferBean refer;

    public String getCode() {
        return code;
    }
    public void setCode(String code) {
        this.code = code;
    }
    public String getUpdateTime() {
        return updateTime;
    }
    public void setUpdateTime(String updateTime) {
        this.updateTime = updateTime;
    }
    public String getFxLink() {
        return fxLink;
    }
    public void setFxLink(String fxLink) {
        this.fxLink = fxLink;
    }
    public NowBean getNow() {
        return now;
    }
    public void setNow(NowBean now) {
        this.now = now;
    }
    public ReferBean getRefer() {
        return refer;
    }
    public void setRefer(ReferBean refer) {
        this.refer = refer;
    }
    public static class NowBean {
        private String obsTime;
        private String temp;
        private String feelsLike;
        private String icon;
        private String text;
        private String wind360;
        private String windDir;
        private String windScale;
        private String windSpeed;
        private String humidity;
        private String precip;
        private String pressure;
        private String vis;
```

```java
    private String cloud;
    private String dew;

    public String getObsTime() {
        return obsTime;
    }
    public void setObsTime(String obsTime) {
        this.obsTime = obsTime;
    }
    public String getTemp() {
        return temp;
    }
    public void setTemp(String temp) {
        this.temp = temp;
    }
    public String getFeelsLike() {
        return feelsLike;
    }
    public void setFeelsLike(String feelsLike) {
        this.feelsLike = feelsLike;
    }
    public String getIcon() {
        return icon;
    }
    public void setIcon(String icon) {
        this.icon = icon;
    }
    public String getText() {
        return text;
    }
    public void setText(String text) {
        this.text = text;
    }
    public String getWind360() {
        return wind360;
    }
    public void setWind360(String wind360) {
        this.wind360 = wind360;
    }
    public String getWindDir() {
        return windDir;
    }
    public void setWindDir(String windDir) {
        this.windDir = windDir;
    }
    public String getWindScale() {
        return windScale;
    }
    public void setWindScale(String windScale) {
        this.windScale = windScale;
    }
    public String getWindSpeed() {
        return windSpeed;
```

```java
        }
        public void setWindSpeed(String windSpeed) {
            this.windSpeed = windSpeed;
        }
        public String getHumidity() {
            return humidity;
        }
        public void setHumidity(String humidity) {
            this.humidity = humidity;
        }
        public String getPrecip() {
            return precip;
        }
        public void setPrecip(String precip) {
            this.precip = precip;
        }
        public String getPressure() {
            return pressure;
        }
        public void setPressure(String pressure) {
            this.pressure = pressure;
        }
        public String getVis() {
            return vis;
        }
        public void setVis(String vis) {
            this.vis = vis;
        }
        public String getCloud() {
            return cloud;
        }
        public void setCloud(String cloud) {
            this.cloud = cloud;
        }
        public String getDew() {
            return dew;
        }
        public void setDew(String dew) {
            this.dew = dew;
        }
    }

    public static class ReferBean {
        private List<String> sources;
        private List<String> license;

        public List<String> getSources() {
            return sources;
        }
        public void setSources(List<String> sources) {
            this.sources = sources;
        }
        public List<String> getLicense() {
```

```
            return license;
        }
        public void setLicense(List<String> license) {
            this.license = license;
        }
    }
}
```

根据天气预报的请求地址创建描述网络请求的接口 RequestWeatherInfo，其具体代码如下：

```
import retrofit2.Call;
import retrofit2.http.GET;
import retrofit2.http.Query;

public interface RequestWeatherInfo {
    @GET("now")
    Call<WeatherInfo> getWeatherInfobyCityID(@Query("location") String cityID, @Query("key") String key);
}
```

由请求地址可知使用 GET 方式提交 HTTP 请求，请求的子路径为 now。请求时调用的方法 getWeatherInfobyCityID()将会返回一个城市的天气信息对象。请求方法的两个参数对应请求地址问号"?"后的两个查询参数 location 和 key，其中，查询参数 location 应为所要查询城市的代码，而查询参数 key 对应在和风天气网申请到的天气预报服务的密钥。这两个查询参数使用@Query 注解来表示，其类型均为字符串类型。注意，在调用时需传入自己所申请的服务密钥 KEY。

新建 WeatherActivity 页面，采用相对布局，添加用于显示天气图标的 ImageView 控件，以及用于显示天气信息的一些 TextView 控件。通过 Retrofit 创建请求对象，发起异步请求即可在页面上显示请求到的城市的天气信息了，其 WeatherActivity 类的具体实现代码如下所示：

```
import androidx.annotation.NonNull;
import androidx.annotation.Nullable;
import androidx.appcompat.app.AppCompatActivity;
import android.graphics.drawable.Drawable;
import android.os.Bundle;
import android.util.Log;
import android.widget.ImageView;
import android.widget.RelativeLayout;
import android.widget.TextView;
import com.bumptech.glide.Glide;
import com.bumptech.glide.request.target.SimpleTarget;
import com.bumptech.glide.request.transition.Transition;
import retrofit2.Call;
import retrofit2.Callback;
import retrofit2.Response;
import retrofit2.Retrofit;
```

```java
import retrofit2.converter.gson.GsonConverterFactory;

public class WeatherActivity extends AppCompatActivity {
    private TextView textViewName;
    private TextView textViewWeather;
    private TextView textViewTemp;
    private ImageView imageView;
    private TextView textViewDir;
    private TextView textViewSpeed;
    private TextView textViewHumi;
    private Retrofit retrofit;
    private RequestWeatherInfo requestWeatherInfo;
    private WeatherInfo weatherInfo;
    private RelativeLayout relativeLayout;

    @Override
    protected void onCreate(Bundle savedInstanceState) {
        super.onCreate(savedInstanceState);
        setContentView(R.layout.activity_weather);
        textViewName = findViewById(R.id.textViewName);
        textViewWeather = findViewById(R.id.textViewWeather);
        textViewTemp = findViewById(R.id.textViewTemp);
        imageView = findViewById(R.id.imageView);
        textViewDir = findViewById(R.id.textViewDir);
        textViewSpeed = findViewById(R.id.textViewSpeed);
        textViewHumi = findViewById(R.id.textViewHumi);
        relativeLayout = findViewById(R.id.relativeLayout);
        retrofit = new Retrofit.Builder()
                .baseUrl("https://devapi.qweather.com/v7/weather/")
                .addConverterFactory(GsonConverterFactory.create())
                .build();
        String cityID = getIntent().getStringExtra("cityID");
        textViewName.setText(getIntent().getStringExtra("cityName"));
        requestWeatherInfobyCityID(cityID);
    }

    private void requestWeatherInfobyCityID(String cityId)
    {
        requestWeatherInfo = retrofit.create(RequestWeatherInfo.class);
        Call < WeatherInfo > call = requestWeatherInfo.getWeatherInfobyCityID(cityId,
"               ");        //这里需传入自己所申请的服务密钥 KEY
        call.enqueue(new Callback < WeatherInfo >() {
            @Override
            public void onResponse(Call < WeatherInfo > call,
Response < WeatherInfo > response) {
                WeatherInfo weatherInfo = response.body();
                textViewWeather.setText(weatherInfo.getNow().getText());
                textViewTemp.setText(weatherInfo.getNow().getTemp() + "度");
                textViewDir.setText("风向:" + weatherInfo.getNow().getWindDir());
                textViewSpeed.setText("风速:" + weatherInfo.getNow().getWindSpeed());
                textViewHumi.setText("湿度:" + weatherInfo.getNow().getHumidity());
                if(weatherInfo.getNow().getText().equals("阴"))
                {
```

```
Glide.with(WeatherActivity.this).load("https://rg.itsven.cn/static/img/clouds.png").into
(imageView);
Glide.with(WeatherActivity.this).load("https://rg.itsven.cn/static/img/bz1.jpg")
                        .into(new SimpleTarget<Drawable>() {
                            @Override
                            public void onResourceReady(@NonNull
Drawable resource, @Nullable Transition<? super Drawable> transition) {
                                relativeLayout.setBackground(resource);
                            }
                        });
                }
                if(weatherInfo.getNow().getText().equals("晴"))
                {
Glide.with(WeatherActivity.this).load("https://rg.itsven.cn/static/img/sun.png").into
(imageView);
Glide.with(WeatherActivity.this).load("https://rg.itsven.cn/static/img/bz2.png")
                        .into(new SimpleTarget<Drawable>() {
                            @Override
                            public void onResourceReady(@NonNull
Drawable resource, @Nullable Transition<? super Drawable> transition) {
                                relativeLayout.setBackground(resource);
                            }
                        });
                }
            }
            @Override
            public void onFailure(Call<WeatherInfo> call, Throwable t) {
                Log.e("CALLweatherInfo", "onFailure: " + t.getMessage() );
            }
        });
    }
}
```

其天气信息页面 WeatherActivity 的显示效果如图 10.18 所示。

图 10.18　天气信息页面 WeatherActivity 的显示效果

10.4 小结

本章详细地讲解了 Android 的网络编程，包括使用 WebView 控件浏览网页、HTTP、JSON 数据格式的解析、使用 Retrofit 框架发起 HTTP 异步请求等知识。在实际开发中，手机应用程序往往要访问网络资源并会对互联网中的 JSON 数据进行解析，因此希望读者能够熟练掌握本章中使用 Retrofit 框架通过 HTTP 获取互联网资源的开发技巧。

习题 10

1. 简述 HTTP 的工作过程。
2. 简述如何使用 Gson 类库进行 JSON 数据的解析，并对如下 JSON 字符串进行解析：

```
{
    "name":"中国",
    "province":[{
        "name":"黑龙江",
        "cities":{
            "city":["哈尔滨","大庆"]
        }
    },{
        "name":"广东",
        "cities":{
            "city":["广州","深圳","珠海"]
        }
    },{
        "name":"台湾",
        "cities":{
            "city":["台北","高雄"]
        }
    },{
        "name":"新疆",
        "cities":{
            "city":["乌鲁木齐"]
        }
    }]
}
```

3. 简述使用 Retrofit 框架通过 HTTP 获取互联网资源的过程。
4. 尝试自己开发一个 SpringBoot 的项目与 Android 应用程序通过 Retrofit 框架进行交互。

第11章 综合项目——科学饮食管理系统

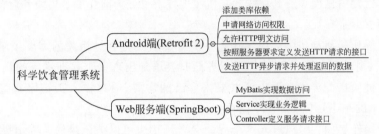

主要内容

- 了解科学饮食管理系统功能模块。
- 理解科学饮食管理系统架构。
- 掌握科学饮食管理系统开发流程。

难点

Android 端与 Web 服务端之间的交互。

随着生活质量的不断提升人们对营养与健康的关注度越来越高，科学饮食管理可以帮助人们控制日常营养摄入，所以设计一款优秀的科学饮食管理 App 很有必要。

关于 Android 科学饮食管理系统，移动端使用 Android Studio 开发工具、Retrofit 2 框架和 SQLite 数据库；Web 服务端使用 IDEA 开发工具、SpringBoot 框架和 MySQL 数据库。

11.1 科学饮食管理系统简介

科学饮食管理系统的主要任务是可以根据用户的需要提供特定方向的膳食营养素摄入记录的需要，实际生活中，合理的营养膳食的摄入可以有效改善人们的健康，以食疗养，因此开发出来一款手机 App 应用，随时随地将自己膳食营养素摄入量信息记录下来。

对于个体而言，某种营养素的需要量是机体为处于并维持其良好的健康状况，在一定的时期必须平均每天吸收该营养素的最低量（也称生理需要量）。由于个体对某种营养素的需要量受到年龄、性别、生理特点、劳动状况等诸多因素的影响，即便是同一个体特征也会由于生理的差异而导致对食物和营养素的需要量存在显著的生物学差异，因此无法提出一个适用于人群中所有个体的营养素的需要量。科学饮食管理系统的目的正是对个体在营养素参考摄入量方面有针对性地指导,以防止人们由于长期营养素不足或过量带来的可能性危险。

科学饮食管理系统内置一份建议健康食谱,可以供用户进行查询;选择营养饮食类别,记录自己营养素摄入量,选择标注时间,输入对应备注进行特殊条目显示;通过DRIs按日查询、DRIs按月查询、DRIs按年查询,更加清晰明了地了解自己的营养素摄入量情况;通过数学计算类目占比,规范个人饮食习惯。

11.2 功能模块设计

科学饮食管理系统的功能主要涉及水果营养信息模块、蔬菜营养信息模块、食谱营养信息模块以及营养饮食信息模块。其功能的具体实现模块可分为前台展示模块和后台管理模块。

前台展示模块:

(1) 导航操作的滑动进入:用户在第一次安装软件时会对相关的需要了解的操作进行自主滑动学习,滑动到最后一张操作学习界面时会增加出现一个立即进入的跳转按钮。

(2) 搜索健康搭配界面:用户可以根据自己的需要,通过输入关键字来查看对应详细信息,主要显示的是不可以一起食用的相关信息。

(3) 记录操作界面:用户进入这个界面会看到相应可以被记录的营养值摄入量。

(4) 更多操作界面:点击之后可以跳转到搜索相应备注显示条目界面、设置时间显示相应条目界面以及显示相应年份、月份、日的总摄入量。

后台管理模块:

(1) 用户信息:管理注册登录的用户信息。

(2) DRIs营养素摄入量记录统计:对记账的用户DRIs营养素摄入量进行按日统计、按月统计和按年统计。

其功能结构如图11.1所示。

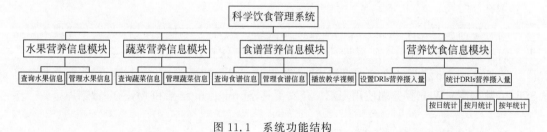

图11.1 系统功能结构

11.3 数据库设计

数据库设计在整个项目开发的过程中扮演着重要的角色,不全面考虑设计数据库会影响到项目内置功能以及相应界面显示内容的数据填充,数据库的设计不可省略。

11.3.1 数据库实体

数据库实体是指数据库管理系统中用于数据管理而设定的各种数据管理对象,这些对象中所存储的数据也就是数据库实体。数据库实体相应会封装关于健康食谱的相关数组、表示可记账类型的表、表示记账内容的表等。

（1）用户的实体包括 ID、用户名、用户密码、用户类型、登录状态，属性如图 11.2 所示。

图 11.2　用户实体属性

（2）水果营养信息实体包括 ID、水果名称、水果营养信息、水果图片，属性如图 11.3 所示。

（3）蔬菜营养信息实体包括 ID、蔬菜名称、蔬菜营养信息、蔬菜图片，属性如图 11.4 所示。

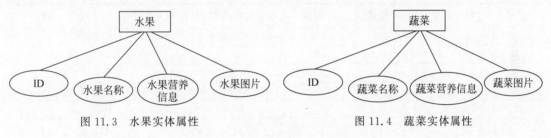

图 11.3　水果实体属性　　　　　图 11.4　蔬菜实体属性

（4）食谱实体包括 ID、食谱名称、食谱营养信息、食谱图片、食谱视频，属性如图 11.5 所示。

图 11.5　食谱实体属性

（5）营养饮食实体包括 ID、日期、DRIs 名称、DRIs 摄入量、用户 ID，属性如图 11.6 所示。

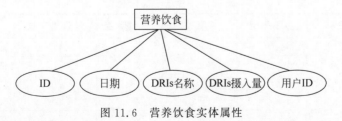

图 11.6　营养饮食实体属性

11.3.2　数据库表设计

本项目使用 SQLite 轻型数据库，设计了相关信息表，各表结构如表 11.1～表 11.5

所示。

（1）用户信息表：用来存放用户的信息，包括 ID、用户名、用户密码、用户类型、登录状态 5 个字段，如表 11.1 所示。

表 11.1 用户信息表

序号	字段名称	字段类型	长度	是否允许为空	是否为主键	说明
1	id	integer	11	No	Yes	编号自增
2	username	varchar	20	No	No	用户名
3	userpassword	varchar	20	No	No	用户密码
4	usertype	varchar	20	No	No	用户类型
5	loginstate	integer	2	No	No	登录状态

（2）水果营养信息表：用来存放不同水果详细内容的相关信息，包括 ID、水果名称、水果营养信息、水果图片、备注 5 个字段，如表 11.2 所示。

表 11.2 水果营养信息表

序号	字段名称	字段类型	长度	是否允许为空	是否为主键	说明
1	id	integer	11	No	Yes	编号自增
2	fruitname	varchar	20	No	No	水果名称
3	fruitdetail	varchar	1200	No	No	水果营养信息
4	fruitimg	varchar	100	No	No	水果图片
5	comment	varchar	500	No	No	备注

（3）蔬菜营养信息表：用来存放不同蔬菜详细内容的相关信息，包括 ID、蔬菜名称、蔬菜营养信息、蔬菜图片、备注 5 个字段，如表 11.3 所示。

表 11.3 蔬菜营养信息表

序号	字段名称	字段类型	长度	是否允许为空	是否为主键	说明
1	id	integer	11	No	Yes	编号自增
2	vegetname	varchar	20	No	No	蔬菜名称
3	vegetdetail	varchar	1200	No	No	蔬菜营养信息
4	vegetimg	varchar	100	No	No	蔬菜图片
5	comment	varchar	500	No	No	备注

（4）食谱营养信息表：用来存放不同食谱详细内容的相关信息，包括 ID、食谱名称、食谱营养信息、食谱图片、食谱视频、备注 6 个字段，如表 11.4 所示。

表 11.4 食谱营养信息表

序号	字段名称	字段类型	长度	是否允许为空	是否为主键	说明
1	id	integer	11	No	Yes	编号自增
2	recipename	varchar	20	No	No	食谱名称
3	recipedetail	varchar	1200	No	No	食谱营养信息
4	recipeimg	varchar	100	No	No	食谱图片
5	recipevideo	varchar	100	No	No	食谱视频
6	comment	varchar	500	No	No	备注

（5）营养饮食信息表：用来记录用户每日的营养 DRIs 摄入值情况，包括 ID、日期、DRIs 名称、DRIs 摄入量、用户 ID、备注 6 个字段，如表 11.5 所示。

表 11.5 营养饮食信息表

序号	字段名称	字段类型	长度	是否允许为空	是否为主键	说明
1	id	integer	11	No	No	编号自增
2	cdate	integer	10	No	No	日期
3	drisname	varchar	20	No	No	DRIs 名称
4	drisvalue	float	50	No	No	DRIs 摄入量
5	userid	integer	11	No	No	用户 ID
6	comment	varchar	500	No	No	备注

11.4 项目界面显示、操作模块的实现

项目界面导航模块实现了页面导航进入 App 模块、首次安装时的通过导航页面进入应用模块并初始化应用数据、搜索查找健康食谱搭配模块、记录条目模块以及查看相关日期、月份、年份的营养素摄入量记账信息模块与添加营养素摄入量记账信息模块等。

11.4.1 页面导航模块

首次安装，会进入页面导航的滑动界面，如图 11.7 所示。

模块相对界面布局中，有一个可以滑动变换的 ViewPager 控件，页面布局主要体现背景资源图片的不同，之后的线性布局，有水平排列的三个文本显示控件，滑动时文本控件的背景和字体样式会有所变化，线性布局的下面有一个立即进入的按钮，在滑动到最后一个界面背景时才会显示。

图 11.7 滑动运行截图

引入一个数据源集合（简单说下 ViewPager 的 PagerAdapter 的使用方法：返回

ViewPager 页面的个数、判断是否为同一张图片、向 ViewPager 控件添加内容、从 ViewPager 控件移出内容),默认缓存数据源个数。其核心代码如下:

```java
pages.add(view1);
pages.add(view2);
pages.add(view3);
startPageAdapter = new StartPageAdapter(pages);
viewPager.setAdapter(startPageAdapter);
```

初始化 ViewPager 页面资源,根据数组资源,控制需要改变的背景,添加到 ViewPager 数据源集合中,创建适配器对象。

通过 ViewPager 控件的导航,点击最后一页的"开始体验"按钮进入系统的主页面。其核心代码如下:

```java
buttonStart.setOnClickListener(new View.OnClickListener() {
    @Override
    public void onClick(View v) {
        SharedPreferences sp = getSharedPreferences("userInfo", MODE_PRIVATE);
        String name = sp.getString("name", "#");
        if (name.equals("#")) {
            initData();
        }
        Intent intent = new Intent(MainActivity.this, LoginActivity.class);
        startActivity(intent);
    }
});
```

11.4.2 登录界面模块

这个模块的界面主要设计在一个相对布局当中,相对布局的背景是项目内部资源中的一张图片,相对布局里有一个 ImageView 控件(用来显示用户的一个头像)和两个 EditText 控件(一个用于用户输入用户名,另一个用于用户输入密码,第一次登录时会使用用户输入的用户名和密码进行自动注册,该用户名和密码会通过 SharedPreferences 保存在本地项目路径下的 userInfo.xml 文件中。SharedPreferences 是 Android 平台上一个轻量型存储类,可以持久化存储少量数据,首先调用获取数据实例对象的方法,然后调用编辑获取数据对象方法)。保存用户信息的核心代码如下:

```java
private void saveUser()
{
    SharedPreferences sp = getSharedPreferences("userInfo",MODE_PRIVATE);
    SharedPreferences.Editor editor = sp.edit();
    editor.putString("name",editTextName.getText(
)
.toString(
)
);
    editor.putString("pwd",editTextPwd.getText(
)
```

```
        .toString(
  )
);
    editor.apply();
    Toast.makeText(this,
"用户信息保存成功!",Toast.LENGTH_SHORT).show();
}
```

读取用户信息的核心代码如下:

```
private void readUser()
{
    SharedPreferences sp = getSharedPreferences("userInfo",MODE_PRIVATE);
    String name = sp.getString("name","");
    String pwd = sp.getString("pwd","");
    editTextName.setText(name);
    editTextPwd.setText(pwd);
}
```

"登录"按钮通过 Intent 转至系统主界面。登录界面如图 11.8 所示。

图 11.8　登录界面

11.4.3　科学饮食管理系统主界面

该模块显示一个垂直方向的线性布局,最上面是一个水平线性布局,包裹了 4 个 Button 控件,分别是水果、蔬菜、食谱和营养饮食,通过这 4 个按钮可进入相应的模块。在该水平线性布局的下方仍然是一个水平线性布局,包括了一个文本输入框控件 EditText 和一个"搜索"按钮控件 buttonQuery,文本框可以输入要查找的项目名称,点击"搜索"按钮可进行查找。搜索时可以模糊搜索食物名称，显示相应简略食物信息。搜索框内容不能为空。搜索会将包含输入框信息内容的食物名称全部显示出来，内部默认数据源清空，填入新的集合数据源时，提示适配器更新。执行刷新操作时会将搜索框内容重置，恢复显示默认数据源（调用全部数据方法，将全部数据保存起来，把全部数据添加到内部数据源）。

搜索功能界面之下是一个 ListView 控件，将食物名称、食物略图、食物详情按钮分别写成 Food 类存储到对象，将对象存储到集合，返回集合，也就是 ListView 控件的内部数据源，集合泛型就是定义好的食物类（食物类图片资源 ID 是整型）。

创建 ListView 控件的数据适配器（该适配器继承自 ArrayAdapter < Food >），声明 ListView 控件所在的界面，传入内部数据源（数据适配器有 4 个方法，获取 Item 条目的总

图 11.9　科学饮食管理系统主界面

数、根据位置获取某个 Item 对象、根据位置获取某个条目 ID、获取相应位置对应的 Item 视图），在 ListView 只会区域滑动显示一定条目。为了减少内存损耗、泄漏，这里采用 itemView 复用旧视图，创建适配器内部类，封装条目有关控件：一个 ImageView 控件（显示食物略图）、一个 TextView 控件（显示食物名称）和一个 Button 控件（转向食物详情页面）。

界面代码调用 Food 类获取全部数据的方法，把全部的数据添加到内部数据源当中，为 ListView 控件设置适配器，ListView 控件里的条目要设置单向点击监听，获取当前位置的对象数据向详情界面进行传递。

条目单向点击监听跳转到的界面是食物搭配的详情界面，食物类可以被序列化，详情界面从上级界面接收信息显示在相关控件上，将接收到的序列化数据强转为食物类对象，而详情界面的界面布局包裹在可以滚动的 ScrollView 控件中，ScrollView 中只能有一个控件，把需要显示的内容全部加载到一个线性布局里面。科学饮食管理系统主界面如图 11.9 所示。

11.4.4　水果营养信息模块

该模块是进入系统主界面后默认显示的模块，主要是通过一个 ListView 控件显示 SQLite 数据库中的水果营养信息，其数据表为 fruittable。fruittable 数据表中的数据操作通过类 FruitOpenHelper 来进行，该类继承自 Android SDK 中的 SQLiteOpenHelper 类，主要封装了插入水果营养信息数据、按水果名称查询水果营养信息数据、按水果名称更新水果营养信息数据、按水果名称删除水果营养信息数据、获取所有水果营养信息数据等功能。ListView 控件通过 ArrayAdapter 加载所有水果信息数据，点击每个分项中的"详情"按钮跳转至水果详情页面，并通过 Intent 中的附件将该项水果名称传至详情页面，详情页面根据水果名称查询水果营养信息数据显示在详情页面，其效果如图 11.10 所示。

根据水果名称查询水果营养信息数据的代码如下：

```
public String querybyName(String name)
{
        SQLiteDatabase sqLiteDatabase = getReadableDatabase();
        Cursor cursor = sqLiteDatabase.query(tableName,new
String[]{"detail"},"name = ?",new String[]{name},null,null,"name");
        if(cursor.getCount()> 0)
        {
        cursor.moveToFirst();
        String detail = cursor.getString(0);
        return detail;
        }
        else
        {
            return "";
        }
}
```

在水果营养信息的详情页面可添加新的水果营养信息,也可对当前的水果信息进行更新和删除。其更新的页面如图 11.11 所示。

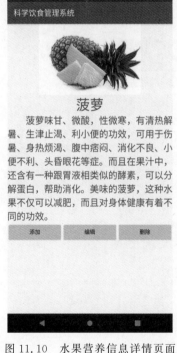

图 11.10　水果营养信息详情页面　　图 11.11　水果营养信息更新页面

更新水果营养信息数据的代码如下:

```
public void updatebyName(String name,String detail)
{
        SQLiteDatabase sqLiteDatabase = getWritableDatabase();
        ContentValues contentValues = new ContentValues();
        contentValues.put("detail",detail);
```

```
            int count = sqLiteDatabase.update(tableName,contentValues,"name = ?",new
String[]{name});
            if(count > 0) Toast.makeText(context,"更新完成!",
Toast.LENGTH_SHORT).show();
        }
```

11.4.5　蔬菜营养信息模块

点击主页上方的"蔬菜"按钮就可以显示蔬菜营养信息模块,该模块的布局与水果营养信息模块的布局基本一致,也是通过 ListView 控件进行展示的,数据来自蔬菜信息表 vegettable,VegetOpenHelper 类通过查询 vegettable 表,将所有蔬菜的相关信息数据封装到 ArrayList＜Veget＞的数组集合中。使用该数据集合中的数据生成 vegetArrayAdapter 对象,将 vegetArrayAdapter 对象设置给 ListView 控件进行显示。其核心代码如下:

```
vegetDbHelper = new VegetOpenHelper(FoodsActivity.this);
vegets = vegetDbHelper.getAllContacts();
vegetArrayAdapter = new VegetArrayAdapter(FoodsActivity.this, vegets);
listView.setAdapter(vegetArrayAdapter);
```

蔬菜营养信息模块实现的效果如图 11.12 所示。

图 11.12　蔬菜营养信息模块

点击蔬菜每个分项中的"详情"按钮同样会跳转至蔬菜详情页面,并可在蔬菜详情页面对蔬菜的营养信息进行添加、更新和删除操作。删除操作会通过蔬菜的名称查找到相应的数据记录,在 vegettable 表中将该数据删除。其核心代码如下:

```java
public void delbyName(String name)
{
    SQLiteDatabase sqLiteDatabase = getWritableDatabase();
    int count = sqLiteDatabase.delete(vegettable,"name = ?",new String[]{name});
    if(count > 0)
    {
        Toast.makeText(context,"删除成功!",Toast.LENGTH_SHORT).show();
    }
}
```

删除记录前会弹出对话框让用户确实后进行删除,对话框通过 AlertDialog 进行实现,设置"确定"与"取消"两个按钮,分别添加点击事件的处理方法,如果确定则调用删除的方法进行删除,删除后返回系统主页面;如果取消则关闭删除对话框。其核心代码如下:

```java
AlertDialog dialog = new AlertDialog.Builder(FoodDetailActivity.this)
        .setIcon(R.mipmap.ic_launcher)//设置标题的图片
        .setTitle(
"删除对话框")//设置对话框的标题
        .setMessage(
"确定要删除此食物信息吗?")//设置对话框的内容
        //设置对话框的按钮
        .setNegativeButton(
"取消", new DialogInterface.OnClickListener() {
            @Override
            public void onClick(DialogInterface dialog, int which) {
                dialog.dismiss();
            }
        })
        .setPositiveButton(
"确定", new DialogInterface.OnClickListener() {
            @Override
            public void onClick(DialogInterface dialog, int which) {
                vegetDbHelper.delbyName(food.getName(
)
);
                dialog.dismiss();
                finish();
            }
        }).create();
dialog.show();
```

蔬菜营养信息详情页面中删除信息的实现效果图,如图 11.13 所示。

11.4.6 食谱营养信息模块

点击主页上方的"食谱"按钮就可以显示食谱营养信息模块,该模块也是通过 ListView 控件进行展示的,数据来自食谱信息表 recipetable。点击食谱的"详情"按钮可以跳转至食谱营养信息的详情页面,该页面采用了一个垂直方向的线性布局,线性布局上依次是一个 ImageView 控件用于显示食谱的图片;一个 TextView 控件用于显示食谱的名称;一个 TextView 控件用户显示食谱的详情;一个视频播放的按钮可以跳转至该食谱制作的教学视频；一行进行添加、编辑、删除的按钮。其实现效果如图 11.14 所示。

图 11.13　删除当前蔬菜信息　　　图 11.14　食谱营养信息的详情页面

点击"播放视频"按钮跳转至该食谱制作的教学视频，跳转时通过 Intent 的附件将该视频的相关信息传递给播放视频的页面，播放视频的页面会根据该食谱在数据表 recipetable 中 recipevideo 字段的值加载食谱并进行播放。其核心代码如下：

```
recipe = (Recipe) intent.getSerializableExtra("recipe");
mediaController = new MediaController(this);
videoPath = "android.resource://" + getPackageName() + "/" + recipe.get VideoPath ();
videoView.setVideoPath(videoPath);
videoView.setMediaController(mediaController);
mediaController.setMediaPlayer(videoView);
videoView.start();
```

播放食谱制作教学视频的实现效果如图 11.15 所示。

11.4.7　搜索食物营养信息模块

点击主页上方的"搜索"图标，会跳转到根据食物名称查找显示相关记录条目功能界面，主要显示一个显示规定条目的 ListView 控件。界面代码声明了相关食物记录条目数据的集合数据源和一个在显示记录条目界面代码中使用过的 ListView 数据适配器，数据适配器需要注意声明有关食物对象的数据集，数据适配器主要匹配显示到的信息。

模块界面代码需要先将显示搜索内容的区域显示为没有搜索内容，提示搜索框中的内容不能为空，相关食物查找到的数据源调用了数据库访问类 FoodDbHelper 中的根据名称显示相关条目的方法 querybyName(String name)。其核心代码如下：

图 11.15　播放食谱制作教学视频

```
boolean isExist = false;
if(foods!= null)
{
    for(Food food:foods)
    {
        if(food.getName(
)
.equals(
editTextQuery.getText(
)
.toString(
)
.trim(
)
)
)
        {
            isExist = true;
            Intent intent = new Intent(FoodsActivity.this,FoodDetailActivity.class);
            intent.putExtra("food",food);
            startActivity(intent);
        }
    }
    if(!isExist)
    {
        Toast.makeText(FoodsActivity.this,
"抱歉,没有找到!",Toast.LENGTH_SHORT).show();
    }
}
```

11.4.8 营养饮食信息模块

点击主页上方的"营养饮食"按钮就可以进入营养饮食信息模块,该模块通过 ListView 控件进行展示,主要分为营养饮食概述、DRIs 简介、DRIs 设置、DRIs 按日查询、DRIs 按月查询、DRIs 按年查询 6 个栏目。

该模块各栏目中的数据与后台 DRIs 的记账 Web 服务进行交互获取。本系统 Android 应用与后台 Web 服务进行交互的 HTTP 请求的发起采用 Retrofit 2 框架进行,首先需要在 Gradle 配置脚本中添加 Retrofit 2 框架类库的依赖。其类库依赖的具体配置如下:

```
implementation 'com.google.code.gson:gson:2.2.4'
implementation 'com.squareup.retrofit2:retrofit:2.9.0'
implementation 'com.squareup.retrofit2:converter-gson:2.0.2'
```

Retrofit 2 框架类库的版本采用 2.9.0 的版本,用于 JSON 与类对象转换的 Gson 类库版本为其 Retrofit 2 框架内的为 2.0.2 的版本,通用的 JSON 转换类库版本为 Google 的 Gson 2.2.4 版本。

其次需要在 App 的项目配置文档 AndroidManifest.xml 中配置申请获取访问互联网资源的访问授权。其具体的授权申请如下:

```
<uses-permission android:name="android.permission.ACCESS_WIFI_STATE" />
<uses-permission android:name="android.permission.INTERNET" />
<uses-permission android:name="android.permission.ACCESS_NETWORK_STATE" />
```

因所请求的 Web 服务在本地进行调试运行,故使用 localhost 地址为服务器地址,其具体请求的主地址为 http://localhost:8081/account/,这里 Web 服务器使用 Tomcat 服务器,所设置的端口号为 8081,而 Web 服务的项目路径为 account。

在营养饮食信息模块中,DRIs 设置栏目可设置某日某次的 DRIs 摄入值,此 DRIs 摄入值记录数据通过 POST 请求发送给 Web 服务。其请求定义的接口信息如下:

```
@POST("add")
Call<BaseResponse<Account>> add(@Body Account account);
```

Web 服务返回的基本响应数据中包含请求状态码 StatusCode,如果请求状态码 StatusCode 的值为 0,则请求成功,数据成功提交到服务端并保存到服务器中的 MySQL 数据库中,对应的数据表为营养饮食信息表 diettable。

在界面中点击"确定"按钮发起 POST 请求,请求通过异步的方式发起,Web 服务器一旦返回信息则通过接口回调的方式进行异步处理。异步请求的处理接口 Callback 中定义了两个方法,其中 onResponse(Call call, Response response)方法表示响应成功的处理过程,这里请求的响应结果包含在参数 Response 的 Body 中,可通过 response.body()方法获得;而 onFailure(Call call, Throwable t)方法则表示响应失败的处理过程,其请求失败的信息包含在参数 Throwable 的 Message 中,可通过 t.getMessage()方法获得。

如果用户提交 DRIs 摄入量信息成功则为用户显示添加成功,否则在后台输出失败的调试信息。其核心代码如下:

```
Call < BaseResponse < Account >> call = requestDayList.add(account);
call.enqueue(new Callback < BaseResponse < Account >>() {
    @Override
    public void onResponse(Call < BaseResponse < Account >> call,
Response < BaseResponse < Account >> response) {
        if(response.isSuccessful()&&response.body()!= null) {
        Toast.makeText(AddDRIActivity.this, "添加成功", Toast.LENGTH_SHORT).show();
        //finish();}
        }
        @Override
        public void onFailure(Call < BaseResponse < Account >> call, Throwable t) {
        Log.d("err",t.getMessage());}
    });
```

设置 DRIs 摄入量的实现效果如图 11.16 所示。

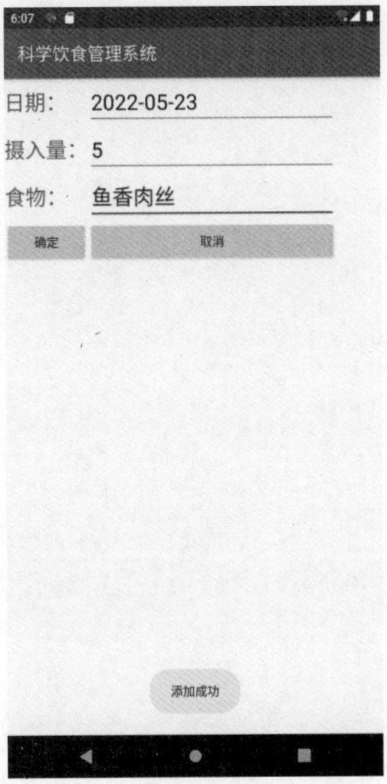

图 11.16　设置 DRIs 摄入量

11.4.9　DRIs 统计查询模块

在营养饮食信息模块中，DRIs 营养素摄入量的统计查询可以分为按日查询、按月查询和按年查询三个栏目，分别对当前用户 DRIs 某一天总的摄入值、某一个月份总的摄入量或某一年份总的摄入量进行统计查询，其具体统计查询的过程由后台的 Web 服务根据该用户的 DRIs 营养素摄入量进行具体统计反馈，Android 端仅提交查询统计的 HTTP 请求，请求以 GET 请求的方式发起。其具体的请求定义接口信息如下：

```
//按日查询
@GET("daylist")
```

```
Call < DayList > getDayList();
//按月查询
@GET("monthlist")
Call < Monthlist > getMonthlist();
//按年查询
@GET("yearlist")
Call < Yearlist > getYearlist();
```

所有请求返回相应数据的实体类对象。

以按日查询为例,其请求获取的实体类为 DayList,除包含返回的网络基本状态信息外,主要包含了该用户某一日期总的摄入量。其 DayList 具体代码如下:

```
public class DayList {
    /**
     * code : 0
     * msg : 成功
     */
    private int code;
    private String msg;
    private List < DataBean > data;
    public List < DataBean > getData() {
        return data;
    }
    public static class DataBean {
        /**
         * cdate : 2020 - 01 - 10
         * je : 10.0
         */
        private String cdate;
        private double je;
        public String getCdate() {
            return cdate;
        }
        public double getJe() {
            return je;
        }
    }
}
```

DRIs 营养素摄入量的统计查询请求同样采用异步的方式发起,返回的统计数据包含在 Response 参数的 Body 中,可通过 response.body().getData()方法获取,遍历 dataBeanList 集合中的所有数据记录,将其转换为字符串的数组 ArrayList < String >中,同时累加总的摄入量,将其封装到 DietNote 对象中,通过 Intent 传递到 DietShowActivity 页面进行显示。其核心代码如下:

```
Call < DayList > call = requestDayList.getDayList();
    call.enqueue(new Callback < DayList >() {
        @Override
```

```java
            public void onResponse(Call<DayList> call, Response<DayList> response) {
                //textView.setText(response.body().getMsg());
                List<DayList.DataBean> dataBeanList = response.body().getData();
                ArrayList<String> strings = new ArrayList<>();
                double total = 0;
                for(DayList.DataBean dataBean:dataBeanList)
                {
                    strings.add(dataBean.getCdate() + "摄入量:" + dataBean.getJe());
                    total = total + dataBean.getJe();
                }
                dietNote = new DietNote();
                dietNote.setTitle("DRIs 按日查询");
                dietNote.setContent("膳食营养素总摄入量" + total);
                dietNote.setStringList(strings);
              Intent intent = new Intent(DietListActivity.this,DietShowActivity.class);
                intent.putExtra("dietnote",dietNote);
                startActivity(intent);
            }

            @Override
            public void onFailure(Call<DayList> call, Throwable t) {
                Log.d("err",t.getMessage());

            }
        });
```

DRIs 营养素摄入量按日查询的统计结果如图 11.17 所示。

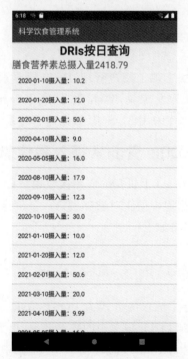

图 11.17　DRIs 营养素摄入量按日查询的统计结果

DRIs 营养素摄入量按月统计、按年统计的 Android 端请求实现与按日统计的请求实现基本相同,其实现的显示效果如图 11.18 所示。

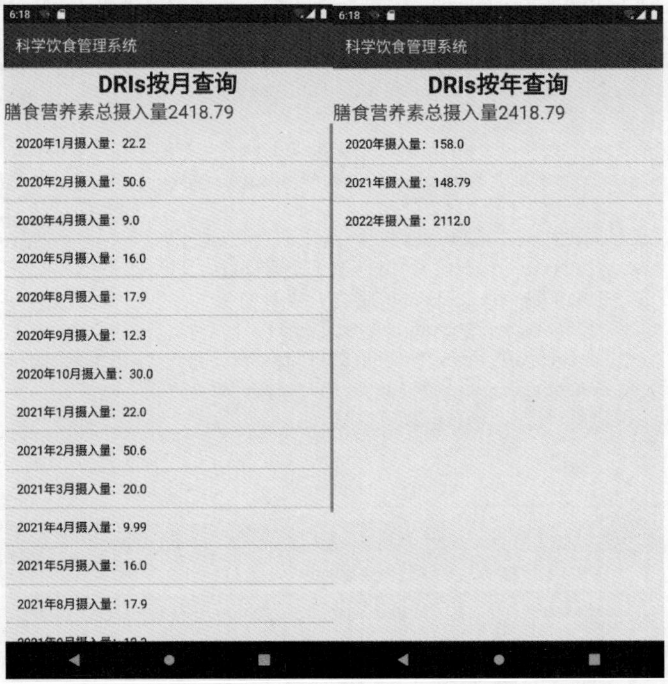

图 11.18　DRIs 营养素摄入量按月与按年查询的统计结果

11.5　科学饮食 Web 服务端的实现

科学饮食 Web 服务端模块实现了用户管理、DRIs 营养素摄入量信息管理等服务。服务端采用 IntelliJ IDEA 2020 开发工具,Web 服务采用 SpringBoot 框架,数据访问采用 MyBatis 持久层框架,后台数据库采用 MySQL 数据库管理系统。

11.5.1　保存用户 DRIs 营养素摄入量信息

Web 服务端能够通过 Android 手机端的应用接收用户提交过来的 DRIs 营养素摄入量信息,并将其保存到后台的数据库中便于对营养素摄入量信息的统计分析。

其接收数据请求的接口定义如下:

```
@PostMapping("/add")
public BaseResponse add(@RequestBody Account account);
```

@PostMapping 注解规定了请求的方式为 POST 请求,方法的参数中@RequestBody 注解规定了提交的数据封装在 HTTP 报文中的数据部分,提交的类型为实体类 Account 的一个具体的对象。实际运行中 Account 对象在 HTTP 报文中以 JSON 格式进行传输,其与 JSON 格式的转换由 SpringBoot 框架自动转换。请求成功后会返回给前端一个 BaseResponse 的对象,该对象主要包含状态码、描述信息、响应数据(采用泛型表示可以接受通用的数据类型)等属性,用以前端判断提交请求的结果。

服务端接收到前端请求的数据 Account 对象后,调用 MyBatis 持久层框架中的 saveAccount(Account account)方法将其保存在后台数据库中。其在数据映射 Mapper.xml 中所定义的 SQL 语句如下:

```xml
<insert id="saveAccount" parameterType="Account" useGeneratedKeys="true" keyColumn="id" keyProperty="id">
         insert into c_details (cdate, drisname, drisvalue,userid)
         values(#{cdate},#{ drisname },#{ drisvalue },#{ userid })
</insert>
```

11.5.2 统计用户 DRIs 营养素摄入量信息

在 Web 服务端最核心的功能是对用户所提交的营养素摄入量进行统计分析,主要分为按日统计、按月统计和按年统计。其请求的接口定义如下:

```java
@GetMapping("/daylist")
    public BaseResponse daylist();
@GetMapping("/monthlist")
    public BaseResponse monthlist();
@GetMapping("/yearlist")
    public BaseResponse yearlist();
```

@GetMapping 注解规定了请求的方式为 GET 请求,请求成功后会返回给前端一个 BaseResponse 的对象,BaseResponse 的对象的数据部分封装了统计的结果。其按日统计的接口实现代码如下:

```java
@GetMapping("/daylist")
    public BaseResponse daylist() {
        List<Map> maps = accountService.selectAccountByDay();
        BaseResponse retMsg = new BaseResponse(StatusCode.Success);
        retMsg.setData(maps);
        return retMsg;
    }
```

营养素摄入量统计在映射 Mapper.xml 中所定义的 SQL 语句如下:

```xml
<select id="selectAccountByDay" resultType="map">
select cdate,sum(drisvalue) as je from c_details group by cdate order by cdate
</select>
<select id="selectAccountByMonth" resultType="map">
select year(cdate) as year,month(cdate) as month,sum(drisvalue) as je from c_details group by year(cdate),month(cdate) order by year(cdate) ,month(cdate)
</select>
<select id="selectAccountByYear" resultType="map">
select year(cdate) as year,sum(drisvalue) as je from c_details group by year(cdate) order by year(cdate)
</select>
```

11.6 小结

本系统重点实现的是日常 DRIs 营养素摄入量记账功能。通过前后端的交互能够按日统计、按月统计和按年统计，可以看到有记录的每天总的摄入量、每月总的摄入量和每年总的摄入量，可以根据特定时期总的摄入量调整自己的饮食习惯，这对以后进行相应健康饮食规划有很大帮助。

参 考 文 献

[1] 欧阳燊.Android Studio 开发实战:从零基础到 App 上线[M].3 版.北京:清华大学出版社,2022.
[2] 李季,张雨,李航.Android 移动应用开发教程(微课版)[M].北京:人民邮电出版社,2023.
[3] 张思民.Android Studio 应用程序设计(微课视频版)[M].3 版.北京:清华大学出版社,2023.
[4] KRISTIN M.Android 编程权威指南[M].4 版.北京:人民邮电出版社,2021.
[5] 安辉.Android App 开发从入门到精通[M].北京:清华大学出版社,2018.
[6] BUDI K.Java 和 Android 开发学习指南[M].2 版.北京:人民邮电出版社,2023.
[7] 赵克玲,吕怀莲.Android Studio 程序设计案例教程(微课版)[M].2 版.北京:清华大学出版社,2023.
[8] 黑马程序员.Android 企业级项目实战教程[M].北京:清华大学出版社,2018.
[9] 黑马程序员.Android 移动应用基础教程(Android Studio)[M].2 版.北京:中国铁道出版社,2019.
[10] 黑马程序员.Android 移动开发基础案例教程[M].2 版.北京:人民邮电出版社,2021.
[11] 李刚.疯狂 Android 讲义[M].4 版.北京:电子工业出版社出版,2019.
[12] 郭霖.第一行代码:Android[M].3 版.北京:人民邮电出版社,2020.

图书资源支持

感谢您一直以来对清华版图书的支持和爱护。为了配合本书的使用,本书提供配套的资源,有需求的读者请扫描下方的"书圈"微信公众号二维码,在图书专区下载,也可以拨打电话或发送电子邮件咨询。

如果您在使用本书的过程中遇到了什么问题,或者有相关图书出版计划,也请您发邮件告诉我们,以便我们更好地为您服务。

我们的联系方式:

清华大学出版社计算机与信息分社网站:https://www.shuimushuhui.com/

地　　址:北京市海淀区双清路学研大厦 A 座 714

邮　　编:100084

电　　话:010-83470236　010-83470237

客服邮箱:2301891038@qq.com

QQ:2301891038(请写明您的单位和姓名)

资源下载: 关注公众号"书圈"下载配套资源。

书圈

清华计算机学堂

观看课程直播